ROUNDING UP THE USUAL SUSPECTS?

For Nicola, Joanna, Ellen and Robert

Captain Louis Renault (Claude Rains):
'Major Strasser has been shot. Round up the usual suspects.'

Casablanca (Dir. Michael Curtiz, Warner, 1942)

Rounding Up the Usual Suspects?

Developments in contemporary law enforcement intelligence

PETER GILL
Liverpool John Moores University

Ashgate

Aldershot • Burlington USA • Singapore • Sydney

Published by
Ashgate Publishing Ltd
Gower House
Croft Road
Aldershot
Hants GU11 3HR
England

Ashgate Publishing Company
131 Main Street
Burlington
Vermont 05401
USA

Ashgate website: http://www.ashgate.com

British Library Cataloguing in Publication Data
Gill, Peter, 1947-
    Rounding up the usual suspects? : developments in
    contemporary law enforcement intelligence
    1. Law enforcement - Great Britain 2. Intelligence service -
    Great Britain 3. Law enforcement - North America
    4. Intelligence service - North America
    I. Title
    363.2'52'0941

Library of Congress Control Number: 00-134475

ISBN 1 84014 923 X

Printed and bound by Athenaeum Press, Ltd.,
Gateshead, Tyne & Wear.

Contents

List of Figures

List of Tables

Preface

Asked by a colleague some years ago what I was working on, I replied, 'A book about police intelligence'. He laughed heartily: 'But, surely, that's an oxymoron - like military music'. A lively discussion ensued both as to whether policing could be 'intelligent' and, if so, was this not likely to be repressive?   I hope that the book provides at least partial answers to these and other questions that emerge - can policing, more often associated with 'doing' than with 'thinking', re-invent itself in the Information Age?   By appropriating new technological tools, is policing changing or is it merely using them in its historical mission of controlling the 'dangerous classes'?

Since 1993 police in the UK have been developing the idea of 'intelligence-led policing' and seeking to shift organisational practices in order to attain their goals more effectively.   Charting and explaining the progress, or otherwise, of this shift is the central aim of this study. While concerned mainly with developments in the UK, it seeks to compare these with police intelligence structures in North America, specifically, Canada and New York State. The value of comparative analysis is that it provides a broader context for evaluating police practices than the study of any single country can do.   The developments described here are on-going, for example, the UK Government has just introduced to Parliament a Regulation of Investigatory Practices Bill - I have not tried to deal systematically with events after February 2000.

Many of the people I have talked to and interviewed during the research and writing of this book would prefer not to be named but I would like to thank them all for their help. My thanks to the Canadian High Commission in London and John Moores University who awarded me research grants that enabled the research to be carried out in, respectively, Canada and New York.   Thanks to Pete Ackerley for encouraging the project and arranging access to North Wales Police. I am grateful to Alana Barton and Roger Evans for giving me access to the unpublished data from their study of Merseyside Police.   Several people have read parts of the book - Alana Barton, Blair Southerden and Steve Tombs, and gave me useful comments. Geoff Ponton kindly read and provided valuable editorial suggestions on the whole thing.   Reg Whitaker has provided much assistance through his regular circulation to myself and other colleagues of the results of his trawlings on the Net. The meetings with colleagues in the ESRC Seminar Series on 'Policy Responses to Transnational Organised Crime' have been constantly stimulating, in particular, I am grateful to Adam Edwards and Peter Klerks for their suggestions. Thanks to colleagues in the Centre for Criminal Justice at Liverpool John Moores University for the congenial environment they provide and some more specific feedback on a paper I delivered to a research seminar. Thanks to i2 for giving me permission to reproduce the Figures that appear in chapter nine. As ever, none of these named or unnamed helpers are responsible for any errors or the particular interpretations that appear.

Pete Gill, May 2000

Finally, most thanks to Pen, Ell and Rob who have all been very supportive throughout this project.

List of Abbreviations

ACC	Assistant Chief Constable (UK)
ACIIS	Automated Criminal Intelligence Information System (Canada)
ACPO	Association of Chief Police Officers (UK)
ADA	assistant district attorney (US)
BATF	Bureau of Alcohol, Tobacco and Firearms (US)
BCU	Basic Command Unit
BOSS	Bureau of Special Services (New York)
CACP	Canadian Association of Chiefs of Police
CLEU	Co-ordinated Law Enforcement Unit (Canada)
CIA	Central Intelligence Agency (US)
CID	Criminal Investigation Department (UK)
CISC	Criminal Intelligence Service Canada
CISO	Criminal Intelligence Service Ontario (Canada)
CIU	criminal intelligence unit
CPA	crime pattern analysis
CPIA	Criminal Procedure and Investigations Act 1996 (UK)
CPS	Crown Prosecution Service (UK)
CSIS	Canadian Security Intelligence Service
DA	district attorney (US)
DC	Detective Constable (UK)
DCI	Detective Chief Inspector (UK)
DI	Detective Inspector (UK)
DOI	Department of Investigation (New York)
DS	Detective Sergeant (UK)
DEA	Drug Enforcement Administration (US)
FBI	Federal Bureau of Investigation (US)
ECHR	European Convention on (or, Court of) Human Rights
EEA	European Economic Area
EU	European Union
FinCEN	Financial Crimes Enforcement Network (US)

FIO	Field Intelligence Officer (UK)
FIB	Force Intelligence Bureau (UK)
FRU	Field Reconnaissance (sometimes Research) Unit (Northern Ireland)
GCHQ	Government Communications Headquarters (UK)
GIS	Geographical Information Systems
GRID	General Reference Information Database (UK Customs)
HIDTA	High Intensity Drug Trafficking Area (US)
HMIC	Her Majesty's Inspectors of Constabulary (UK)
HOLMES	Home Office Large Major Enquiry System (UK)
IALEIA	International Association of Law Enforcement Intelligence Analysts
ICT	Information and Communication Technologies
ICU	Intelligence Co-ordination Unit (UK Customs)
IOCA	Interception of Communications Act 1985 (UK)
ILA	International Longshoremen's Association (US)
ILETS	International Law Enforcement Telecommunications Seminar
ILP	intelligence-led policing
INS	Immigration and Naturalization Service (US)
IRS	Internal Revenue Service (US)
JIC	Joint Intelligence Committee (UK)
LEIU	Law Enforcement Intelligence Unit (US)
LIO	Local Intelligence Officer (UK)
MI5	Security Service - internal security agency (UK)
MI6	Secret Intelligence Service - foreign intelligence agency UK)
MOU	Memorandum of Understanding
MTA	Metropolitan Transportation Authority (New York)
MTP	Metro Toronto Police (Canada)
NAFTA	North American Free Trade Area
NCIS	National Criminal Intelligence Service (UK)
NCS	National Crime Squad (UK)
NDIU	National Drugs Intelligence Unit (UK)
NEGIS	North East Gang Information System (US)
NID	National Intelligence Division (UK Customs)
NPM	new public management

NYPD	New York City Police Department
NYSOCTF	New York State Organized Crime Taskforce
NYSP	New York State Police
NYSPIN	New York State Police Information Network
OCCSSA	Omnibus Crime Control and Safe Streets Act 1968 (US)
OPP	Ontario Provincial Police (Canada)
OSNP	Office of the Special Narcotics Prosecutor (New York)
PACE	Police and Criminal Evidence Act 1984 (UK)
PCA	Police Complaints Authority (UK)
PII	Public Interest Immunity (UK)
PIRA	Provisional Irish Republican Army
PITO	Police Information Technology Organisation (UK)
PNC	Police National Computer (UK)
PRI	*Partido Revolucionario Institucional* (Mexico)
RCMP	Royal Canadian Mounted Police
RCIO	Regional Criminal Intelligence Office (UK)
RCS	Regional Crime Squad (UK)
RICO	Racketeer Influenced and Corrupt Organizations Act (US)
RISS	Regional Information Sharing Systems (US and Canada)
RUC	Royal Ulster Constabulary (Northern Ireland)
SAS	Special Air Service (UK)
SCIS	Secure Criminal Information System (Canada)
SIS	Secret Intelligence Service - also MI6 (UK)
SNARE	Statewide Narcotics Apprehension and Reporting Effort (New York)
SNIP	Statewide Narcotics Indexing Program (New York)
TCG	Tasking and Co-ordination Group (UK)
TSU	Technical Support Unit (UK)
UDA	Ulster Defence Association (Northern Ireland)

1 Governance, Information and Police

Introduction

This study starts from the premise that writing and research on the police has been over-weighted towards issues of police action compared with their information and intelligence activities. This can be explained primarily by the more immediate and dramatic impact on the public's consciousness of, say, police abuse of their powers of arrest or of 'reasonable force'. Accordingly, researchers have concentrated far more on the patrol and public order functions of police. Thus the academic analysis of policing has 'mirrored' the 'cop culture' (Reiner, 1992, 107-37) that emphasises action and excitement over more prosaic activities such as information gathering and analysis. Indeed, these are apt to be discredited by police themselves as of low priority compared with enforcement (Manning & Hawkins, 1989, 146; Martens, 1990, 4-5). If police surveillance has been researched then it has tended to be in the narrower sense of a concern with the propriety of information-gathering practices, often in the context of 'political' policing (e.g. Gill, 1994; Turk, 1982). There are few if any references to 'information' or 'intelligence' in general books on policing in the UK (an exception is Kinsey, 1986) and the primary literature relating to police use of information and intelligence has been written largely by and for practitioners in North America (Andrews & Peterson, 1990; Godfrey & Harris, 1971; Peterson, 1995). When research has addressed the issue of intelligence it has tended to be because the research interest is in a specific area of policing where intelligence work is central, for example, 'organised crime' (Beare, 1996) or transnational co-operation (Anderson *et al*, 1995) or some specific aspect of the intelligence process is the subject, for example, the use of computers (Campbell & Connor, 1986) or informants (Dunnighan & Norris, 1996). Where research has been carried out more generally into the police's use of information then the findings suggest that it is not very effective: as Tremblay and Rochon have argued, public policing is currently organised in such a way that police remain under-informed about social

1

expectations of them, overloaded with unanalysed crime data and inadequately committed to strategic research (1991, 269-83).

It is important to redress this imbalance for a number of reasons: first, because policing as a multi-dimensional activity is, actually, largely concerned with the processing of information rather than 'crime control' or 'law enforcement' *actions*:

> Information is a critical feature of modern societies and is the essential and central feature of policing. (Manning, 1992, 352)

The notion of 'policing by consent' by civilian officers armed with few more powers than the ordinary citizen always rested crucially on the notion that they could gather information concerning crime in local areas. Indeed, the mark of the 'good copper' was the productivity of his sources and ability to turn them into 'collars'.   The other primary police functions - order maintenance and emergency social service - also involve information. A British Crime Survey found that 40% of police-public contacts involve information-sharing (Morgan & Newburn, 1997, 80). Re-visiting two police activity surveys conducted during the 1980s in London and Merseyside, 48% of officers time in the former and 53% in the latter clearly related to information processes (for example, interviewing, paperwork, observation). The three other main activities - patrolling, waiting/refreshments and 'other' (46% in London, 38% in Merseyside) - were not included although some portion of them clearly will also have involved such important information processes as casual conversation and gossip (Kinsey, 1985, 45; Smith, 1983, 38-9). For example, operational officers (both plain clothes and uniformed) are involved in the gathering and dissemination of information *via* incident reports of numerous types far more than they are in making arrests or other enforcement activities.   Officers in support roles are involved in the receipt, processing and dissemination of information both to other units within the police organisation and outside agencies.

Thus the police are, as Ericson and Haggerty, show, 'knowledge workers' (1997, 19-30). Further, they argue that policing is best understood in terms of a model of risk communication in which

> Policing consists of the public police co-ordinating their activities with policing agents in all other institutions to provide a society-wide basis for risk management (governance) and security (guarantees against loss). (1997, 3)

It is undoubtedly the case that public police are now must usefully seen as embedded in extensive security networks – an argument that will be developed in chapter two.   Another reason for paying greater attention to contemporary developments is that law enforcement agencies themselves are placing greater emphasis on explicit intelligence activities.   It is proposed to review briefly these developments and then to develop an analytical model of police intelligence systems as the basis for a critical analysis of these activities.

Researching information and intelligence processes

This remains a difficult area for study: police in general are now more open to researchers but problems remain. The shift during the 1990s from ideas of political to market accountability have affected the way in which police agencies deal with the outside world. While police agencies have always sought to present their preferred image to the public, their response to outsiders perceived as threatening police autonomy, such as local police authorities in the 1980s, was often defensive – chief constables would simply refuse to provide the reports or information requested. Resort to this traditional form of 'information control' – secrecy – remains part of the police armoury, for example, witness the current debate as to the extent of the exemption police will have from future Freedom of Information legislation in the UK, but the shift to market accountability has contributed to a subtle shift in strategy. Now that police are required by a number of central authorities to produce information regarding performance indicators and local policing plans they have lost the ability to control entirely what information they provide. Therefore there is more premium on the way in which that information is provided and so, for example, chief constables' annual reports are now glossier and with a higher ratio of pictures to text than formerly. The same can be seen with other public sector bodies.

Yet, despite this shift, close study will reveal that less hard information upon which police performance might be assessed is actually being produced. The utility of current performance indicators can be criticised (e.g. Loveday, 1999, 356-62), for example, because of their measurement of what *can* rather than what *should* be measured, but also because their construction remains within the control of the police themselves and, as such, is at the mercy of a number of time-honoured practices of fiddling the books. Information produced by agencies is part of broader strategies to 'persuade' people that they perform well.

Therefore research by outsiders still poses a threat to police autonomy and gatekeepers must be negotiated with. Intelligence work is seen as a particularly sensitive area into which outsiders are not entirely welcome. Some of this reluctance is entirely merited in that premature publication may compromise operations or publicity regarding intelligence methods may alert potential targets and, at the most extreme, the identification of informers may endanger their safety. There will therefore always be some tension between what researchers may want to know and what police are prepared to discuss, but unjustifiable defensiveness remains. The particularly secret nature of intelligence operations within law enforcement organisations typically enjoying extensive powers vis-à-vis citizens pose a major challenge to democratic aspirations. Ten years ago James Anderton, then Chief Constable of GMP and President of ACPO, could say 'police intelligence is not a matter for public discussion' (Campbell, 1987) but the new significance given to this area of policing and its continuing propensity to scandal and abuse, especially the increased use of informants, demand much more significant and systematic research.

The research process consisted of the following main stages: the collection of the relevant published literature, interviewing over seventy practitioners working at various levels of the intelligence process and collecting such official documentation as could be made available. In Canada access was attained eventually to all the agencies approached but in both New York and the UK some declined. Certainly, police now receive a large number of requests for access and information from researchers and it is simply not feasible for them to respond positively to all requests. However, the overriding question that agencies apparently ask themselves when approached is: 'What's in it for us?' and the opportunities for gatekeepers to intervene, both formally and informally, are extensive. For example, following correspondence with the UK ACPO Crime and Research Committees, the author was informed by letter that this research had been 'approved' and forces were being circulated accordingly. However, a few weeks later, while negotiating access with a particular police force, he was informed that approval had been withdrawn. After further inquiries, it appeared that this reversal had been determined by a particular chief constable, chair of one of the ACPO committees who, after further thought, decided there was nothing in the research for police. He did not see fit to communicate or explain this decision to the author.

However, even if gatekeepers are successfully passed, primary reliance on official accounts can result in a lop-sided view but greater resources than were available here would be required to obtain the counter-balance of the perceptions of those who are the targets or consumers of police intelligence. But even that, as Margaret Beare points out, would not solve entirely the intrinsic difficulties of researching areas in which most of the 'players' prefer to maintain secrecy *vis-à-vis* outsiders (1996, 22-36).

Even after access has been gained, the interests of practitioners and researchers do not necessarily coincide and this can give rise to tensions, for example, the incompatibility between organisational demands on practitioners for 'results' (often in the short rather than the long term) measured in seizures or arrests and those on academics for the publication of medium to longer term analyses of 'the problem'. But this tension may be creative; if academic study is to retain its credibility and autonomy it must retain a critical edge - it must scrutinise the terms of official discourse not just adopt them. After all, official definitions are not forged in a neutral scientific way but, rather, in the crucible of the bureaucratic politics characterising modern states. Andrew Sanders has criticised criminal justice research in Britain for its failure specifically to stand outside categories defined *for* the intellectual rather than *by* the intellectual (1997, 187, emphasis in original).

Meanwhile, academics may be criticised by politicians requiring (re-) election and their advisers who clearly need to promise 'solutions' to problems and thus to reassure insecure publics that 'something is being done'. But researchers need not be defensive in the face of official scepticism: some 'policy problems' are so intractable that it would be highly misleading for academics to endorse the 'quick-fixes' that are often demanded by politicians and officials (Dintino & Martens, 1983, 102). Also, they can observe that closed policy-making circles are not noted for their success compared with those open to critical scrutiny and assistance to policy in the longer term can only be given by work that does not simply accept as axiomatic prevailing definitions of the terms of debate.

Yet, even though academic work need not be directly oriented to the production of policy-outcomes, it should be communicated in a manner that is meaningful and relevant to practitioners as well as to students and other researchers. If academics want to fulfil their intellectual duty of 'speaking truth to power' (Said, 1994, 63-76) then they are obligated to make themselves understood. But if academics are to remain sceptical of 'power' then it must be borne in mind that 'public' or 'state' power is not the only kind of power there is; as soon as one accepts that crime is not entirely the figment of official imaginations then it has to be acknowledged that considerable power is wielded 'private' criminal enterprises.

There is another twist to the task facing academics when analysing current police intelligence practices: strategic criminal intelligence is nothing more than applied social science research (Dintino & Martens, 1983, 100) and, to the extent that practitioners actually make use of the concepts and methods of social science in their work, outside researchers will sometimes have the sense of looking into a mirror. This is potentially confusing but if it induces some critical self-reflection it is not a bad thing.

Contemporary governance

Before embarking on a detailed examination of police information and intelligence processes, it is important to establish a broader framework of contemporary governance that provides the context for these processes. Recent theoretical work seeks to explain contemporary politics and society in terms of processes of 'surveillance' (Dandeker, 1990, 119-33; Giddens, 1985, 181-92; Lyon, 1994). This is not simply the use by police and other agencies (private and public) of specific information-gathering practices; rather it has two primary components: first, the gathering and storing of information, second, the supervision and monitoring of people's behaviour. The second component is sometimes referred to as 'disciplinary practices' (Foucault, 1983) and reflects the attempt to influence (or 'control' or 'steer') behaviour rather than just observe it. Nicola Lacey suggests that 'social ordering practice' is a better way to conceive of criminal justice because it avoids the narrower instrumentalism of words such as 'control' and can incorporate the important symbolic aspects also (Lacey, 1994, 28-34).

Contemporary notions of governance crucially depart from traditional sovereign notions of command; instead they are centred upon a 'process of co-ordination, steering, influencing and balancing' the interactions of public and private groups (Kooiman, 1993, 255) or what Andrew Dunsire calls 'collibration': government 'participating in the conflict of forces' (Dunsire, 1993, 34). On the face of it this seems reminiscent of classical pluralist ideas of state policy simply being the outcome of the interaction of competing organised interests, but the crucial difference here is that the state is an active player, it has its own will and preferences and seeks to re-shape the 'environment' to some extent in its own image. This is closer to 'neo-pluralism' in which democratic representation is down-graded as the state becomes 'professionalised' (Dunleavy & O'Leary, 1987, 288-318).

But the state may not be successful: some early commentators on the growth of privatisation in policing saw it as one among a number of examples of 'rule at a distance' but later came to recognise that the development of disparate sites of power could frustrate state preferences (Shearing, 1996, 302-3 fn4). This is consistent with other post-structural views of power that emphasise the inadequacy of examining the contemporary exercise of power by means of traditional sovereign models; instead they propose that power must be viewed in terms of its practices or 'micro-techniques' which normalise individuals and groups through forms of surveillance in a variety of personal, technical, bureaucratic and legal forms (Clegg, 1989, 191-2). Foucault's particular characterisation of this replacement for sovereign state forms was 'governmentality': the ensemble of institutions, procedures, strategies and tactics that exercise power over the population primarily by means of apparatuses of security. What distinguishes this from earlier sovereign notions is that rather than obedience to law being an end in itself, the laws themselves are simply one among a number of tactics to be employed by government that is aimed at an end which is 'convenient' for the things and people who are governed (Foucault, 1991, 95-104). Foucault, like the pluralists, was more interested in the means of government that in whose interests it was conducted. Thus there could be a plurality of specific aims, rather than any notion of the imposition of some 'common good'.

Governance and markets

This study is concerned with the specific issue of police intelligence processes regarding crime. Crucial insights here can be derived from political science and economics, and their inter-relation. Economists are centrally concerned with the operation of markets and have, in comparison with other social sciences, developed highly sophisticated models of their operations. These models are criticised on the grounds that they make unrealistic assumptions about the rationality of people's behaviour but it is impossible for any social scientist to avoid making some such assumptions without resorting to explanations based on some mixture of madness and magic. Without entering the debate as to how far models based on rational choice can be taken in analysing areas such as crime, we might accept that:

to state that eminently economic variables have only negligible effects on most individual agents' choices concerning illegal activities seems to be at odds with a growing body of empirical evidence. (Fiorentini & Peltzman, 1995, 2)

Thus it is not necessary to accept the restricted and controversial view that 'criminals' are fully informed and rationally-deciding individuals to acknowledge the broader wisdom of viewing many criminal phenomena as occurring within markets whether they are officially recorded or not. Over 90% of recorded - mainly 'street' -crime in the UK involves 'property' rather than personal offences: if the proceeds are for personal consumption then this represents an alternative to purchase in a legal market; but if the proceeds are sold on then they enter a broader illegal market. The term 'grey markets' is used increasingly to describe that space in which legal shades into illegal (e.g. Karp, 1994). Much other crime is not reported to police for various reasons – yet is also clearly related to the issue of markets: for example, corporate crime is clearly concerned with the corporations' market share and profitability (Slapper & Tombs, 1999, 141-49). Occupational crime, to take another example, is concerned with the activities of individuals or groups within employment whose motives vary – they may simply seize the opportunity provided by the job for personal acquisition either at the expense of consumers or employers or may see the opportunity for crime as part of action within a labour market. The other main area of unreported crime is so-called 'victimless' crime – drugs, prostitution and so on – in which the provision of some goods and services under certain circumstances has been outlawed and the marketplace for them is consequently illegal.

If it is accepted that much crime is, at least in part, an economic phenomenon then it is relatively easy to establish that what is described as 'organised crime' is even more centrally concerned with markets. This issue will be discussed in more detail in chapter three but, for now, suffice to say that almost all definitions, official or otherwise, refer to the language of profit and markets. In the US the Organized Crime Task Force of the President's Crime Commission began:

The core of organized crime activity is the supplying of illegal goods and services – gambling, loan-sharking, narcotics, and other forms of vice – to countless numbers of citizen customers. (Fiorentini & Peltzman, 1995, 3)

The more recent definition adopted by the UK National Criminal Intelligence Service (NCIS) and in turn derived from Interpol is:

> Organised crime constitutes any enterprise, or group of persons, engaged in continuing illegal activities which has as its primary purpose the generation of profits, irrespective of national boundaries. (Clay, 1998, 94)

Given these definitions, it is not difficult to see why criminal organisations may well be compared – formally by economists or fictionally by novelists – as 'firms'.

Nor, turning from the issue of crime itself to the attempt to control it in some fashion, should it be difficult to establish the relevance of political science (though it remains disappointing that so few political scientists have given the issue attention). First, and most obviously, the process by which markets come to be officially determined through law to be either 'legal' or 'illegal' is political. Second, there is the crucial relationship between the state and market-regulation. Formally, if a market is defined as 'illegal' then the state seeks its suppression, whereas if it is defined as 'legal' then it may seek to regulate it, to a greater or lesser extent depending on the nature of the goods and services traded. For example, traders in some products will require a license - alcohol, explosives – while in others no license is required although there may be periodic inspections in relation to health and/or safety. However, beyond the formal level, it is more useful to view the interaction of state and markets as occupying a spectrum in which the precise nature of the relation in any one market is defined only partially by the nature of the law and to a greater extent by the interaction between the organisations and individuals occupying key positions within the state or the market place (see further discussion below).

Third, there is the related but broader issue of legitimacy. Notwithstanding the current discussion of a shift from sovereignty towards 'steering' it remains the case that a 'monopoly of legitimate force' remains, for many analysts, the *sine qua non* of the existence of a state. Thus, while states clearly vary enormously in the extent to which they actually do employ violence to ensure obedience with their wishes, all must retain the right to do so in certain circumstances if they are not to be overthrown or wither away into irrelevance. Only at the most formal level can it be said that crime constitutes a challenge to state power (though not in the self-conscious sense of it as part of revolutionary politics). Indeed, crime either by state agents (Friedrichs, 1995, 53-80) or its corporate allies (Snider, 1991) may actually be a constituent of state power.   But to the extent that crime indicates an unwillingness to abide by formal state declarations regarding acceptable behaviour, then it does represent a challenge, though some constitute a greater challenge than others. In general, states will perceive violent crime, even if not directed at state agents, as more demanding of their attention than property crime. This is not to say that all crimes of violence will receive more attention than all crimes of property because issues such as the credibility of victims and the location of the crimes are also involved. But the reason for this general rule is that violence constitutes a more obvious challenge to the state's monopoly than theft or fraud.

This is even more so where the perpetrators of violence are seen as being in some sense 'organised', as evidenced for example by the seriousness of public order offences being determined in part by the numbers involved and the idea of 'common purpose' as a distinguishing characteristic of the most serious, that is, riot.[1] The greatest challenge of all will be seen where some organisation other than the state is seen to be using violence in order to seek compliance with its own 'rules' and commands. If, for whatever reason, criminal organisations achieve a position in which they secure greater allegiance or loyalty than the state then it faces a threat to its ruling monopoly. It is perceptions such as this that have led to the more frequent representation of 'organised crime' as a threat to national security; it remains debatable as to how far this has occurred because of a genuine increase in the criminal threat and how far because those agencies traditionally charged with monitoring national security threats needed a replacement 'enemy' once the Cold War ended (See chapter four).

The difficulties faced by states in meeting these and other challenges - the shift from sovereignty to steering - have been explained partly in terms of the increased 'knowledge' and 'power' problems. The first refers to the impenetrability of social and economic sub-systems in an increasingly complex and diverse world - the state cannot learn how they work. Alternatively, if the state is seeking to control or regulate these sub-systems, the information it requires is as diverse and complex as that possessed by the 'regulated' and there is no way for the state to obtain this information independent of those who are to be regulated.

But even if this knowledge problem can be solved, the power problem arises: the state rarely possesses adequate powers and instruments of policy with which to intervene in the processes of the sub-systems (Francis, 1993, 25-9; Mayntz, 1993, 13-16). Of course, there is much disagreement as to precisely how extensive and new these problems are. Globalisation, for example, is often argued to be the greatest factor reducing states' sovereignty (McGrew, 1992b provides a good summary of the debate). In any event, the extension of surveillance and the creation of an electronic panopticon is the means by which modern states seek to solve these problems, hoping to inculcate self-control while disciplining the recalcitrant. Police intelligence practices are a component of this.

It is quite clear that the state has reduced its commitment across a wide swathe of social and economic policies where privatisation, 'new public management' and 'internal markets' have in various ways replaced earlier ideas of nationalisation, public corporations and central planning. But surely, it will be suggested, it must be different in an area of policy - crime- where states fight hardest of all to maintain the perception that they are in control, where the ability of states to protect public order and the security of citizens is seen as a *sine qua non* of their claim to power? Indeed, this tension can be seen in the classic juxtaposition of *The Free Economy and the Strong State* (Gamble, 1988) under Thatcher in the UK when reducing the state's role vis-à-vis the economy was necessarily buttressed by increased powers and resources for police as they sought to contain the social consequences of economic policies.

Policing is indeed an area of action in which the state cannot give up its 'monopoly of legitimate violence' as an essential badge of its sovereignty. The notion of local 'security networks' in which the crime-oriented and reactive role of police is demarcated from the safety and preventive orientation of other agencies in the network is an interesting example of an attempt to resolve this problem of governance (Johnston, 2000, 167-75). There are times when 'government' - activity backed by formal authority (Rosenau, 1992, 4) - will provide a better description of policing than 'governance' - the outcome of interactions between a mixture of state and other actors. Some people still enjoy and employ greater rights than others to kick down people's doors!

Yet, policing is not immune from the shift away from 'government': if these 'knowledge' and 'power' problems apply to *legal* social and economic activities, then they clearly apply in much greater force when the state seeks to control *illegal* social activities. If, for reasons of personal privacy, competitive advantage or tax avoidance, people seek to protect information about their 'legal' lives from the state, then how much greater is the state's 'knowledge' problem when people seek to protect information about activities that are clearly or arguably illegal?

But state officials rarely acknowledge the depth of these problems because, if they do, they may be seen as forfeiting the ability to rule. Therefore, political rhetoric or, as Garland puts it, 'punitive display' (1996, 445) will be replete with 'wars against crime' (whether organised or not), and 'zero tolerance' even if a combination of the public's failure to report crime, the police's decision not to record it all, and failures to detect or prosecute the perpetrator mean that fewer than 3 in 100 offences against individuals and their property in England and Wales result in a criminal conviction or a police caution (Barclay & Tavares, 1999, 29). Even this low figure, of course, *over*-estimates the real impact of the criminal justice system because so much white-collar, corporate and 'victimless' crime is not captured even by victim surveys. For example, the under-reporting of injuries at work is similarly extensive (Tombs, 2000).

What this means is that the gulf between the symbolic terms in which policies are discussed and their real impact is probably greater in the area of crime than in any other area of state policy. Thus, while considering official responses to crime, it is necessary to heed Murray Edelman's argument that political rhetoric seeks to reassure mass publics in apparently clear terms that their insecurities are being addressed while the essential ambiguity of that language, especially legal language, provides the political space within which administrators and practitioners negotiate how, if at all, policies will actually be translated into action (1964, 22-43, 140-1). Once we escape from an essentially rhetorical world in which law enforcers in possession of sovereign state power 'wage war' against, 'defeat' or 'smash' criminal organisations and move on to consider the reality of state action regarding crime, then we find, of course, that the relationship is far more contingent and complex.

What is suggested here is that it is more useful to view police as 'regulators' than 'enforcers'. This reflects the more general shift from government to governance but there are more specific arguments. Of course, there have always been some areas of law enforcement in which a 'compliance' model is seen to have been more significant than an enforcement one. For example, the work contrasting Inland Revenue policies towards tax evaders with that of the Department of Social Security (now Benefits Agency) towards benefit fraud shows the clear class bias in enforcement policy (Cook, 1991). Similarly, regulatory agencies are found to be more likely to prosecute the greater the difference between the social backgrounds of regulators and regulatees (Baldwin *et al*, 1998, 18-19).

More generally, it has been common to make a distinction between police policies (prosecution-based) and those of 'regulatory' agencies (compliance-based). Robert Kagan (1984), for example, has compared the roles of police and regulatory inspectors in the US and shows that they share certain common features such as both enforcing the law, working mainly in the field and, because of their discretion, deploying a variety of enforcement 'styles'. However, he argues that the differences are greater: their differing social functions and the nature of offences and offenders dealt with. However, as Kagan does acknowledge, some forms of policing are closer to the regulatory model than others. The greatest contrast is between regulation and the reactive policing of street crime in which there is little moral ambiguity about offences such as robbery and burglary compared with 'enterprise' crimes in which there is considerable moral ambiguity and police and regulatory models converge.

Furthermore, whereas, for example, Kagan, presents the choice between compliance and enforcement as a function of informal exercises of police discretion, Ericson and Haggerty argue that:

> ...compliance-based law enforcement operates through police communications systems in much more pervasive and fundamental ways. Many of the rules, formats, and technologies that form the police communication system are shaped by the knowledge requirements of external institutions that operate within a compliance-based mode of regulation. (1997, 439)

In one sense this is not new because the main examples of these external institutions are those that have always pursued compliance strategies, for example the financial securities and insurance industries. But Ericson and Haggerty's argument that these rules and formats are increasing significantly the visibility of policing to both superiors and outside institutions and consequently reducing the discretion available to patrol officers (1997, 438) is more original and certainly contrary to other research that has emphasised the low-visibility of policing and police practices of screening and filtering that shapes 'primary data' (Manning, 1992, 371). Arguably Ericson and Haggerty exaggerate the impact of the new formats and rules as a consequence of their concentration on the collation and dissemination of information by police rather than the police's own decisions regarding the targeting of their surveillance and information gathering. To the extent that police retain a high degree of autonomy then other factors need to be considered in accounting for their surveillance policies and practices (taken in the broader sense of monitoring and disciplining) regarding markets. Put another way, is there evidence of a shift towards compliance-based policing even with respect to street-crime? Certainly, Sparrow argues that police, along with environmental regulators and the IRS in the US are all seeking to change agency policies and structures in similar ways, that is, toward a more proactive and targeted mix of compliance and enforcement strategies (1994, 101-29) and the research into Customs and Excise for this study indicates the same ambition.

The actual mix of enforcement styles is crucially influenced by the negotiated relationship between regulators and traders. Figure 1.1 is based on two intersecting spectra, one indicating the theoretical range of enforcement practices from full enforcement to non-enforcement, the other indicating the range of markets from the unambiguously legal to the clearly illegal. The ellipses shown here illustrate the variety of regulatory practices within different markets; they are not claimed to be specifically validated empirically; rather, they should be read as suggestive. Where enforcement involves sanctions, these are more likely to be criminal in the case of illegal markets and administrative or civil in the case of legal markets. However, the right-hand end of 'Prosecution' indicates, say, that small number of cases in which the Health and Safety Executive does prosecute. Similarly, the left-hand end of 'Administrative penalties' can be illustrated by the practice of 'fining' airlines and lorry drivers where they unknowingly transport illegal immigrants. There is no reason for regulators to seek to disrupt unambiguously legal markets; however, for police and customs agencies, this is clearly becoming an increasingly attractive option because of the 'costs' of evidence-gathering and the unpredictability of the outcomes of prosecution. 'Cautioning', either formal or informal, represents the main official alternative to prosecution for police and is broadly comparable to regulators issuing compliance notices in legal markets. 'Licensing and taxing' is represented in varying degrees of formality. Formal regulation is exercised over a number of dangerous products (for example, alcohol, tobacco, petroleum) that may be legally traded. Traders have to buy licenses, are subject to regulations, forms of inspection and the goods sold are frequently subject to specific taxes. Failure to comply with regulations may ultimately lead to a loss of license. By definition, illegal traders cannot be formally licensed as such, but there are a variety of circumstances in which they may actually receive an informal 'license' to trade from local police or other regulators, any 'payoff' that is being made amounting to a form of taxation. Examples of such relations can be found among street traders, local drug dealers or fences, especially where those concerned act as informants for police (see chapter eight). Of course, traders at this margin are particularly vulnerable to having their 'licenses' revoked and being sanctioned.

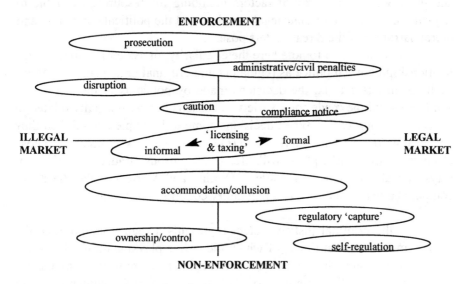

Figure 1.1 Enforcement and Markets

In the bottom half of the continuum, however, traders escape sanctions entirely. In legal markets 'self-regulation' within an industry will be seen as the ideal outcome of effective education and self-control. However, this may not be genuine, reflecting rather a 'creative compliance' that honours the letter rather than the spirit of the rules (Baldwin *et al*, 1998, 20). 'Accommodation and collusion' reflects the fact that all regulatory agencies possess inadequate resources to pursue policies of full enforcement. An inevitable consequence of their selection of priorities (however this is done) is that much illegal trading and regulation-avoidance occurs without any regulatory response. 'Regulatory capture' means that, because of shared ideology and/or personnel, rewards and/or threats, the traders have 'captured' the regulators and ensure non-enforcement. In extreme cases - 'Ownership/control'- the 'regulators' will actually share in the profitability of the market. The apparent involvement of nightclub security staff in dealing drugs for consumption on their premises in various English cities is one illustration of this (World in Action, 1996); the partnership between drugs traffickers and government officials in Mexico (Castells, 1997, 276-86) and Morocco (Werdmolder, 1998) are others. This might occur in legal markets if, say, a regulator was to hold shares in a regulated company. What determines the specific relationship at any particular time and place will be an array of factors including the resources available to regulators, the nature of their legal mandate and the political, economic and moral narration of the threat traders pose.

It is important to note here the circularity, or systemic nature, of the relationship between police as regulators and criminal organisations. As we shall see in chapter six, the decisions made by police as to their priorities and targets will be made for a variety of reasons but are normally related to ideas of 'seriousness'.   When decisions come to be implemented then they will change the criminal marketplace(s) that police seek to control. Those changes in the marketplace will affect the various 'traders' in different ways, not all of them predictable.   Dintino and Martens make the direct comparison between police and regulatory commissions in the US:

> ...the criminal justice system represents a regulatory process, the most severe at times, of which the police play a central role. As is often found in the licit sector, regulatory commissions frequently represent the interests of the more powerful segments of an industry, thereby fostering greater organization and concentrations of power. Police administrators, in their development and execution of Organized Crime control policy may unconsciously (and at times, consciously) encourage and facilitate the organization of marketplaces. (Dintino & Martens, 1983, 56-7)

This encapsulates the central paradox: how far does policing itself create the very conditions it is formally meant to suppress? That police give priority to violence is not surprising since it is the greatest challenge to state legitimacy and is the phenomenon most likely to give rise to public concern with security and safety. 'Monopoly' is a problem for different reasons, primarily ideological. Certainly, within the US one finds that the same ideological beliefs in the superiority of the free market that have inspired government 'trust-busting' since the late nineteenth century exist in the police world and have inspired 'rackets-busting' against the exercise of exploitative monopoly power by criminal organisations. To be sure the American commitment to breaking monopolies has more rhetorical than actual strength, but the commitment remains an important factor in police priority-setting. Thus, Sparrow, for example, suggests that regulatory agencies will want to target the 'worst offenders' because of the damage, if they do not, to their credibility, morale and the message it gives to others (1994, xxvi).

But problems arise because these two objectives - countering monopolies and violence - may frequently be contradictory. As Dintino and Martens suggested above, if police intervene against traders within a competitive market they are likely to succeed through arrest or disruption in driving out of business the smallest or weakest traders and thus actually encourage the growth of a monopoly (1983, 96). When disputes arise between traders in a legal market then there are a number of ways of resolving conflicts including the legal system that are not available in illegal markets and therefore violence is less likely to be used (Dintino & Martens, 1983, 105). What is likely to increase police intervention in a competitive illicit market is the use of violence. If a monopoly has been established then there will be fewer 'turf disputes', less violence and therefore less impetus to police intervention (cf. Fiorentini & Peltzman, 1995, 26).

In considering state action regarding crime another important factor is the nature of the state itself. Specifically, it cannot be assumed that each state is simply a coherent whole pursuing a seamless web of national policy. States may well enunciate very similar-sounding policies regarding crime but when it comes down to the gritty issues of implementation and enforcement then real political interests are at stake. It then becomes clear that states operate at different levels, through different agency and departmental perspectives and that, consequently, policies regarding crime will have to take their chance in competition with other policies whether they be foreign or domestic. We might label this the 'Matrix-Churchill' syndrome: simplifying somewhat, the essentials were that manufacturers sought to sell machine tools equipment to Iraq in the 1980s, an activity that was of interest to a number of state agencies. The Secret Intelligence Service (SIS, a.k.a. MI6) was content to let this trade proceed since the business people involved thereby gained access to secret Iraqi installations and provided the SIS with valuable information; the Department of Trade and Industry, motivated primarily by a desire to maximise profitable exports, connived in avoiding the regulations that were intended to inhibit such sales to certain regimes; and Customs and Excise, with the responsibility for enforcing the export regulations, gathered information about and then prosecuted the traders involved (Scott, 1996). Clearly, there was no single state 'policy' as such in this case.

In other cases the subordination of law enforcement may be even clearer - perhaps the starkest illustration of this has been the US 'war on drugs'. For decades US *anti*-drug policy was subordinated to broader policy aims of anti-communism, for example in Central and South America, Viet Nam and Afghanistan (Block, 1991, 209-26).

Before moving on, let us be clear as to the relationship between law enforcement and illegal enterprises. It has been suggested that it is useful to look at law enforcement as an 'industry' (Walker, 1999, 47-8) since it includes a large number of both private and public agencies, the latter are increasingly imbued with a 'managerialist' ethos and the UK police sought to re-invent themselves as a 'service' through the 1990s.   But this approach can be misleading; for example, precisely who are the 'consumers' of policing? It has long been acknowledged that 'service' functions do occupy a significant proportion of police time but their role as 'social service of last resort' (Punch, 1979) is largely a consequence of their 24-hour availability to offer authoritative interventions in diverse situations, many of which have no criminal justice outcome at all. However, such work also has the least status among police themselves and no amount of managerial rhetoric can conceal the fact that the core mandate of public police is order-maintenance and of private police is surveillance with a view to loss-minimisation.

Given the significance of markets to 'crime', therefore, it is suggested here that the most fruitful perspective is to view police as 'regulators'. The analysis of the interaction of law enforcement and illegal markets will be most enhanced if the former is seen as attempting to establish some form of order or control, that is, governance, over the latter.

Analytical framework: an intelligence system

The factor running through all these developments - more obviously in some cases than in others - is information and intelligence: the knowledge and power or 'governability' problems faced by contemporary states, the centrality of intelligence to police attempts to 'fight' organised crime and the current efforts of police to adapt intelligence techniques to persistent crime problems at whatever level.

Bill Hebenton and Terry Thomas argue in their study of Europe that 'informatization' is the key to understanding and predicting this emergent key practice of policing. This term refers to the 'separate trajectory' of 'police computerisation and the collection, storage and use of personal information' rather than just its usual presentation as a measure aimed at increased efficiency. As such they see it as likely 'to structure further harmonization and integration' of policing systems, 'and perhaps their supranational centralization', citing the UK National Criminal Intelligence Service (NCIS) as an example. Accepting that most police use of intelligence is actually tactical, they argue that nevertheless 'informatization' permits the development of long-term strategic information as a Europeanized resource' (Hebenton & Thomas, 1995, 169-70). Certainly, new information and communication technologies (ICT) *enable* this kind of development; it is a 'power-amplifier' for the police (Nogala, 1995, 193-4).

But then Hebenton and Thomas put the stronger argument that the new technology 'reconstructs' both the aims of policing and the target groups of policework and that these far-reaching changes are being '*brought about* by the pace of technological progress, and the way in which users can see their interests reconfigured around information-based policework' (1995, 170-1). Now, it is certainly the case that there is a developing cadre of police and civilian intelligence specialists but the first part of the statement is too determinist. It can be compared with the kind of technocratic vision maintained by some practitioners, for example, Horst Herold, former *Bundeskriminalamt* President who developed a vision of:

> society-wide data-collations of all crime and justice-related areas, which would enable the police *via* computer to create a self-steering and self-optimizing justice-system. (quoted in Nogala, 1995, 192)

Certainly police intelligence units are one of those areas of policing in which the application of ICT can be seen to be crucial – it is inconceivable that storing, collation, accessing, analysis and dissemination will develop in a context other than that of computerisation. Equally, it is not difficult to identify factors pushing police into greater uses of new technologies, for example, globalisation of markets and crime, expanding 'techno-crimes', continuing professionalisation, the growth of the security-supply industry and the allure of the technological fix (Nogala, 1995, 200-03).

However, technological innovation must be placed within its social context: this includes the patterning effect of policework and the occupational cultures, differentiating between different police roles and the technologies they use. For example, Ackroyd *et al* showed how detectives incorporated ICT into existing police culture and work practices in ways that system designers and senior managers had not intended (1992, 121-41).

There are other factors, for example, the gulf between the promise and the reality of ICT, resource-limitations, system-incompatibilities, the use of counter-measures by criminals, police reluctance to systematise data and public resistance concerning issues of privacy and rights (Manning, 1992, 382-7) - all these may constitute 'counter forces that contradict the straight-forwardness of the process' (Nogala, 1995, 203). Similarly, it might be argued that the somewhat determinist approach of Hebenton and Thomas ignores the key role of analysis in the information process – who is going to ask the questions of the data in order to produce strategic knowledge? There is nothing automatic about this process (See chapter nine).

Thus, in examining police intelligence processes, and the important role of technologies within them, it is important to view them as a dialectic:

> ...the vague notion that available, even conventionally acceptable technology will be used and employed without constraint of police practices and local political traditions is naïve and untenable. (Manning, 1992, 391)

Police will make choices and act as conscious agents, sometimes against the preferences of the purveyors of technology and their own superiors and yet attention must also be given to the fact that while technology may be used to produce and reproduce traditional practices it is slowly modifying them.

Further insight can be gained from cybernetics. 'Kybernaein' (to steer) is the Greek root for both 'cybernetic' and 'govern'. Cybernetics was first comprehensively applied to matters of government over thirty years ago by Karl Deutsch who defined it as 'the systematic study of communication and control in organizations of all kinds' (1966, 76). It certainly represented an advance over then fashionable systems theories of politics which rested on essentially conservative notions of equilibrium because cybernetics is based on the dynamics of information and control systems including the possibility of systems moving from one condition to another (Kooiman, 1993, 37-8). Its concepts are central to the consideration of issues of governance and information control.

For example, 'consciousness' indicates the secondary messages that are attached to primary information within the system (such as circulation lists and assessments of reliability) and which are important in influencing the way in which primary information is dealt with - to the extent that these secondary symbols misrepresent the primary data then the consciousness may be 'false'. 'Will' reflects the point of decision after which conflicting information is blocked; it is a form of 'selective attention' and is closely related to power because hardening a decision is insignificant if there are no facilities for putting it into effect (Deutsch, 1966, 98-100, 110).

'Feedback' is central to the understanding of organisational change: negative feedback leading to some modification, positive to some reinforcement of previous actions. But organisations do not simply react to feedback; their steering and control mechanisms seek to impose the system upon the 'environment' (Deutsch, 1966, 182-99; Kooiman, 1993, 42; Morgan, 1986, 243-5).

In developing an analytical framework that focuses on the information processes within policing, it is important to incorporate key elements of contemporary state and organisational analysis: power, surveillance, governance and steering in order to avoid a naive 'rational' model in which all relevant information is gathered and analyzed before decisions are taken. Rather, the model is one in which there is constant interaction between processes of information and of power. Harry Howe Ransom called for students of intelligence 'to know more about knowledge and power, information and actions' (1980, 148), a request that can be as fruitfully applied to policing as to the more researched areas of foreign and military intelligence.

The objective here is to develop a general model of intelligence systems that can be applied at any level, for example, to specific police criminal intelligence units (CIUs), to police agencies and to policing networks (see chapter two) which represent the interaction of any number of smaller systems. Attempting to understand the consequences of these multiple interactions quickly becomes highly complex but can be characterised as 'loops of mutual causality' (Morgan, 1986, 247-55). Essentially the same point is made in this criticism of the short-sightedness of much policing of organised crime in the US:

> it is extremely important to recognize the intelligence cycle as a continuous process, a critical process which should serve as a tool for law enforcement administrators. Intelligence is designed to guide and shape organized crime control policy and strategy through the continuous probing, questioning, assessing, reassessing, and evaluation of raw data. (Dintino & Martens, 1983, 61)

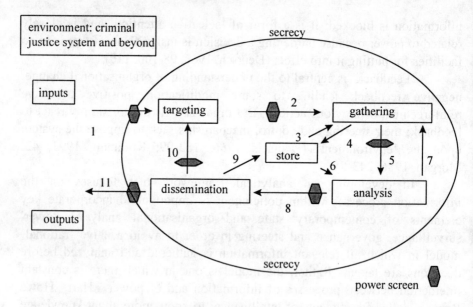

Figure 1.2 Police Intelligence System

Figure 1.2 represents a highly-simplified view of an intelligence system. As in any systems model, the key features are the inputs from its 'environment', the processes by which these are converted within the system into outputs that then 'feedback' into the environment. The notion of system implies a boundary between it and the environment. In intelligence agencies these boundaries may become elaborate procedures for the classification of documents by which the 'gathering' attempts of others are resisted, that is, 'protective security'. (This 'ring' of secrecy is represented graphically in Figure 1.2 and is coincident with the 'boundary' of the intelligence system in the model.) For example, CIUs will often employ yet more extensive security measures than the police agencies to which they belong because of the particular sensitivity of the information they gather. However, in seeking to restrict the flow of information 'need-to-know' procedures may contradict effectiveness and accountability. They may protect privacy and the integrity of future operations but they may also contribute to inadequate analysis, the over-investment of significance in 'secret' information and inadequate dissemination. So where attempts are made to extend intelligence methods into all operational policing at local level the desire to make information available easily to patrol officers may lead to no special security being attached to raw data at all.

Four main stages of the intelligence system ('process' or 'cycle') are normally identified: targeting, gathering, analysis and dissemination. The processes (↑1) by which law enforcement determines its priorities and decides which specific activities, individuals or organisations it is going to target for information gathering (↑2) are discussed in chapter six. Chapter seven discusses the 'store' of information or 'organisational memory' that lies at the heart of any intelligence organisation and which will be consulted (↑3) before other means are deployed to gather information. Technical means of gathering information are also discussed in chapter seven; the use of informants and undercover officers is considered in chapter eight. Some information that is gathered will simply enter the 'store' (↑4) but it should be subject to analysis (↑5). Chapter nine examines analysis, that is, the process of evaluating the meaning of the information gathered and preparing some 'intelligence' product. Analysts may further draw on the store (↑6) or ask for further information to be gathered (↑7). Chapter nine considers also how that product is disseminated (↑8); it may simply be stored (↑9) or provide feedback to those responsible for the initial targeting (↑10) or be passed on to another part of the organisation or another agency (↑11). The discussion here concerns, therefore, the crucial question as to what law enforcement agencies do with intelligence.

Of course, this representation dramatically simplifies what is actually a highly complex and frequently messy process. In applying system and cybernetic concepts to matters of organisational politics, it is important to avoid the trap of providing over-mechanistic accounts of complex political and social realities. A simple systems model might suggest that information processes resemble the uninterrupted technical flow which characterise the well-designed computer system. This would be highly misleading and any model must incorporate those factors that filter, divert and block that flow. This is not to say that managers do not *attempt* to achieve increased automaticity in organisational information systems, for example, the increased use of pre-formatting for incident report forms (Ericson & Haggerty, 1997, 357-87) but they are not always successful. These filters are shown in the Figure as 'power screens' because they represent the operation of power within the system. This manifests itself through both 'agency' and 'structure'. The former is the purposive actions of individuals and groups within the system who may act in such a way as to block, divert or accelerate the passage of information; the latter is the routinised institutions, procedures and practices that both empower and constrain agents and that, over time, may be altered by agents' actions.[2]

Conclusion

Transnational, national and local law enforcement agencies are paying increasing attention to intelligence processes. Since existing scholarship has concentrated on issues of police action rather than information and intelligence, it does not give us an adequate means of understanding and explaining their impact. A cybernetic model of the intelligence system is suggested as providing a systematic basis for the examination of police intelligence processes and their place within the wider context of policing.

If models are to be useful, however, it is not enough that they simply help to organise what is known about their subject but that they also assist in the research process, that is, they are heuristic. What questions, then, are suggested by the application of a cybernetic model to police intelligence processes? For example, are the priorities of police intelligence units driven primarily by external demands or by the internal priorities of the Unit? What are the criteria employed to determine targeting priorities, for example, sums of money, presence of violence or corruption of public life? Do CIUs have the resources to mount their own information gathering operations, or are they dependent on information gathered by others which may be passed on only selectively? Do analysts seek to 'disconfirm' the hypotheses generated by investigators or do they look primarily for 'confirmatory' information as part of the 'case construction' that characterises the adversarial criminal process? (McConville *et al*, 1991, 11-13) In other words, what is the balance between positive and negative feedback processes? Were Dorn *et al* correct in suggesting that the tendency for operational matters to be directed by intelligence priorities would increase? (1992, 148) To the extent that the greater emphasis on intelligence work is increasing the flow (both formal and informal) of information between agencies at different levels (transnational, national and local) and in different sectors (public, hybrid and private), is the democratic deficit of policing growing?

Finally there are more conceptual questions relating to the position of the public police within current patterns of governance. On the face of it, a primary reason for the new emphasis on intelligence is that the police have a 'knowledge problem' in the face of ever-more sophisticated means by which illegal markets proceed or by which otherwise legal markets are manipulated. Yet, key aspects of the working of police intelligence systems suggest that they are more about 'governability problems'. Police believe they know who the criminals are: NCIS, for example, targeted 151 'criminals of major significance' and 2,358 'others subject to NCIS interest' in 1997-98 (NCIS, 1998, 8). Similarly, local intelligence efforts are targeted at 'known criminals'; to this extent police believe that the key problem is less that of lack of knowledge than of governability - a lack of the instruments necessary to intervene in criminal processes (c.f. Mayntz, 1993, 13) including specific kinds of information allied with a capacity to control, disrupt or prosecute those 'known' to be criminals. However, if, because of a lack of negative feedback, the intelligence process works mainly to re-cycle the same people then the knowledge problem may actually be deeper than police believe.

Although, as we have seen, the state resists relinquishing its 'monopoly of legitimate force', how immune is policing from the tendencies to governance found in other policy areas? Is the 'enforcement' of law against 'street crime' (organised or not) really so distinct from the 'regulatory' regimes that exist in other areas? Certainly, given the predominance of definitions of organised crime in terms of markets we should expect that police intelligence units will face similar problems as other agencies involved in the regulation of markets. Thus we might expect to find a variety of practices including selective enforcement, 'negotiation' between regulators and regulated (plea-bargaining in return for information), regulators using their information to 'steer' markets towards or away from competition (especially where competition in illegal markets can be associated with violence) and also intervening (officially or unofficially) through a variety of 'taxing' or 'licensing' arrangements.

Because of the significance of the state's symbolic goal of 'controlling' crime, many of these strategies will only reluctantly if at all be openly admitted. Can the application of intelligence techniques make a real difference or is intelligence-led policing condemned to the failure of other regulatory efforts because of a combination of inadequate resources, impenetrable markets, and collusion with those being 'regulated', subject only to occasional exercises of 'sovereign' enforcement *pour encourager les autres*.

Notes

1   Public Order Act 1986, ss. 1-5. Similarly, one of the criteria for 'serious' crime in which the Security Service is empowered to act in support of law enforcement is 'conduct by a large number of persons in pursuit of a common purpose'. Security Service Act 1996, s.2.

2   A good summary of the structure-agency debate is provided by Hay, 1995, 189-206.

2 Policing Networks

Introduction

In the first chapter we considered the general questions of governance and the extent to which states retain or lose their ability to 'do something' about crime. Before examining specific law enforcement intelligence processes in the UK and North America (see chapter four onwards), the discussion in this chapter focuses more generally on the organisation of law enforcement.

Policing in Canada, United Kingdom and United States

David Bayley suggests there are four benefits of cross-national study of criminal justice: it extends our knowledge of alternative possibilities; it develops more powerful insights into human behaviour - specifically, what effect do police and society have on each other? Third, it increases the likelihood of successful reform and, fourth, we gain a perspective on ourselves as human beings (1999, 6-11). There is more explicit concern here with the first two - since this study focuses on a specific police function the danger inherent in examining just one country is that wrong assumptions might be made about the extent to which experience in other countries would or should be similar. This is particularly the case if the UK is the single country given the resilience of assumptions of superiority about the criminal justice system despite the evidence of its serious shortcomings.

More specifically, one reason for comparing the UK with North America is that, although 'intelligence-led policing' has been the subject of more explicit policy rhetoric in the former, it is arguable that police in the latter have actually been doing it for longer. A major driver behind this development in the UK has been the issue of European integration and the extent to which policing is part of that project (discussed in more detail below). Therefore it is reasonable to ask why this study does not compare the UK more systematically with European police intelligence systems. Practically, the barriers of space (and language) militated against a thorough comparison of European developments as well as those in UK and North

America but use has been made of the excellent Dutch literature on issues of crime and governance.

An important reason for incorporating the North American comparison is that the dominance of the European issue in debate in the UK debate has disguised to some extent the yet broader issue of globalisation. Particularly since the end of the Cold War when the national and international security interests of the remaining superpower - the United States - have been re-oriented, it is important that Europeans remain alert to the prospect that the 'internationalisation' of law enforcement might more accurately be described as 'Americanisation' (Nadelmann, 1993).

One of the dangers of examining some specific policing process comparatively is that one fails to acknowledge the extent to which policing is shaped by the particular countries' social, economic and political circumstances (Mawby, 1999a, 16). This danger is somewhat reduced by the common Anglo-Saxon origins of policing in these three countries (except in Quebec and parts of the US) but the current organisation of policing in the three countries is significantly different and major aspects of this are discussed below. Two apparently incompatible ideas have to be kept in mind: it is clear that crime and disorder manifest themselves primarily *locally* and therefore the particularities of different agencies matter. Yet in recent years law enforcement agencies have come to view 'serious' crime as increasingly *transnational* and to exchange information and techniques with foreign agencies. It has been suggested that this is contributing to some diminution in the differences between agencies (den Boer, 1999, 59) and, indeed, many police do see themselves as part of an international police community. Leon Trotsky is reported to have said: 'There is but one international and that is the police' (van Maanen, 1974, 84).

The development of policing, wherever it is, must be related to broader processes of state and nation-building. For example, Bayley suggests that, other things being equal, more centralised structures will appear where there is greater violent resistance to emergent states. In the US, for example, where an independent state was established before the end of the eighteenth century, local communities continued to be largely self-regulating. The only period in which policing was centralised in the hands of state governors was during Reconstruction in the South following the civil war (Bayley, 1985, 65-72). By contrast, in Canada the Royal Canadian Mounted Police (RCMP) provides a greater degree of centralisation derived from the perception of a need for colonial-style paramilitary policing to pacify the West, natives, political and labour unrest (Mawby, 1999b, 36-7).

In the US the repression and removal of the native nations was conducted by the Army. In the UK relatively peaceful state-building, comparatively small size and cultural homogeneity meant that governments could feel relatively relaxed about the nineteenth century proliferation of small police forces with a high degree of local autonomy. But since then, and at certain key periods such as following the first world war, centralising measures have been taken to reduce the number of forces and increase the central government influence over policy and practices.

Looked at overall, the structure of Canadian policing appears to be at some stage of a process of uneven development. This reflects the presence of a relatively small population (27 million) occupying an extremely large land mass (though they are concentrated heavily within 100 miles of the US border) with a federal structure of government. The 450 police forces in Canada operate currently at either federal, provincial or municipal/regional level. Municipal police developed from 1835 onwards, first in the larger cities and most are found in Ontario and Quebec; currently they range in size from one or two officers to, for example, 4,000 in Metro Toronto. Ontario and Quebec have their own provincial police forces and there is also the Royal Newfoundland Constabulary. Elsewhere, mainly west of Ontario, most policing is carried out by the RCMP under contract to eleven provinces and 197 municipalities. The North West Mounted Police were created in 1873 as a paramilitary force to police what later became Alberta and Saskatchewan and the RCMP was formed through a merger with the Dominion Police Force in 1920. Thus in much of Western Canada the RCMP are *the* public police; in other parts the RCMP provides federal and provincial policing, for example in British Columbia, while in much of Ontario and Quebec it is just a federal force more akin to the Federal Bureau of Investigation (FBI) (Macleod, 1994, 44-56; Marquis, 1994, 24-43).

The Canadian police structure has been classified by Bayley as 'multiple coordinated': RCMP hegemony in much of Canada providing the co-ordination. Although elsewhere the picture is complicated by joint authority between forces, the federal criminal code is uniform. This division of authority is central to any attempt to understand the development in the last thirty years of specialised intelligence units in Canadian police forces. By contrast, policing in the US has been characterised as 'multiple unco-ordinated' (Bayley, 1985, 57-8) since three or four (municipal, county, state, federal) forces might have simultaneous jurisdiction in any area and the basic criminal code is created at the state level. In a few areas this may not be the case: some Native-American reservations quite literally, 'link'

Canada and the US by straddling the border retain a good deal of legal autonomy and their own police departments (Walker, 1999, 55-6).

The sheer number of law enforcement agencies in the US can cause immense difficulties for the student not least because of the debate as to how many agencies there are!   Thirty years ago the President's Commission reported that there were 40,000 (Walker, 1999, 50), in 1985 David Bayley reduced the overall total to 25,000 (Bayley, 1985) and Samuel Walker says there are 18,000 (1999, 50). To a large extent this historical confusion has derived from disagreements about what should be counted, for example, there are a very large number of very small rural police forces and a constantly changing number of agencies established by 'special districts' such as schools, parks, transit companies and universities.

Table 2.1   **Number of US law enforcement agencies and sworn personnel, 1996**

	No. of agencies	No. of sworn officers ('000s)	%age of sworn officers
Federal	(1977) 113+	74.5	10.1
State	49	54.6	7.4
County	3,088	152.9	20.7
Local	13,578	411.0	55.7
Special District	1,316	43.0	5.8
Texas Constables	738	2.0	0.3
Total	18,769	738.0	100.0

Source: + Bayley, 1985; All other figures from Bureau of Justice Statistics at www.ojp.usdoj.gov/bjs/lawenf.htm

The local policing that predominates has its origins in the original settlements of the early seventeenth century - state and federal policing has only really developed during the twentieth. The 'local' category in Table 2.1 includes both the largest and the smallest forces organised in villages, towns or cities. Over half of them employ fewer than ten officers while the 'big six' - New York City, Los Angeles, Chicago, Houston, Philadelphia, Detroit - employ almost 13% of all sworn officers, NYPD being the largest with 38,000 officers in 1997. In rural areas, the County Sheriff was the earliest police and remains most significant. Most sheriffs are elected and provide policing functions, service the courts and maintain county jails; again, there are few very large departments and many small ones (Walker, 1999, 53-54).

All states except Hawaii have a state police and half of them provide general policing while the other half provide just highway patrol. In some states the actual role varies in different parts of the state, for example, on Long Island the New York State Police provide just a highway patrol while in some more rural areas they provide full police services. Many states also have additional specialist state-wide law enforcement agencies such as the New York State Commission of Investigation that deals with corruption investigations in the public sector (see chapter five).

Although a large number of federal agencies have enforcement or regulatory powers most are not general service agencies and do not provide basic protection and investigation. There are seventeen federal agencies that employ 500 or more sworn officers, the largest being the Immigration and Naturalization Service (INS) with 12,000. The FBI and the Federal Bureau of Prisons each have about 11,000 officers and the Drug Enforcement Administration (DEA) about 3,000. (The role of these agencies is discussed in more detail in chapter five.) Evidence relating to the size of the private security sector in the US is even harder to gather than for public policing but it seems that the ratio of private personnel to public is somewhat higher in the US than in the UK. Walker cites a 1991 Justice Department Survey to the effect that 1.5million are employed in private security - roughly twice the number of sworn officers in the public sector (see Table 2.1) - of whom 1.1 million work in guarding or patrol services (Walker, 1999, 57-8).

Although Bayley's characterisation of the US system as uncoordinated appears to be correct, whether the fragmentation of US law enforcement necessarily leads to inconsistent service delivery, crime displacement and competition rather than co-ordination is a major issue. Certainly, examples of all these can be found but the most comprehensive analysis of this issue concluded that systematic co-operation was common among the 80 metropolitan agencies studied and that their experience contradicted the assumption that co-ordination of activities could occur only under a single overarching hierarchy (Ostrom et al, 1978, 324). How this impacts on intelligence activities is a constant theme through this study. Clearly the multiplicity of law enforcement agencies in the US makes efforts at generalisation very hazardous. This is the main reason why this study concentrates on New York for its comparative examples.

It is relatively easy to describe the basic organisational parameters of policing in the UK. There are 43 forces in England and Wales, eight in Scotland and one in Northern Ireland. In addition there are several forces similar to the 'special districts' in the US - the Atomic Energy Authority Constabulary, Ministry of Defence and British Transport Police. The largest UK force is the Metropolitan Police in London with 28,000 sworn officers. Elsewhere in England and Wales the forces are based on counties, though in several cases they cover more than one, and in Scotland they co-exist with the regions on which local government is based. The smallest force in the UK is Warwickshire's 800 but the issue of amalgamations is still more of a live issue than in the US. The Home Secretary took additional powers in 1994 to facilitate this and in Scotland there was talk of the eight forces being reduced to four, three or even one by the new Scottish Parliament when it convened in 1999 (*Scotland on Sunday*, July 26, 1998, 11).

The number of forces in the UK has steadily diminished since their initial founding in boroughs and counties in the mid- to late-nineteenth century but the remaining local autonomy is guarded by both chief constables and their local police authorities (unless, on occasions, it suits either of them to team up with the Home Office against the other). In other ways, the Home Office has successfully sought a measure of standardisation of policy through Circulars and Regulations that would not even be contemplated by the Justice Department in the US but issues of fragmentation still arise. For example, specialist regional and national squads or task forces have been established in both US and UK to deal with issues of 'cross-border' crime and controversies around the co-operation or otherwise of local forces with these squads are endemic.

In the UK in September1999 there were 125,464 police and 53,254 civilians in England and Wales (*The Guardian*, Feb 10, 2000, 9); 14,854 police and 4,642 civilians in Scotland (1998) and in Northern Ireland 8,485 plus 2,982 full-time and 1,324 part-time reservists (1997). A survey of the private security sector about the same time suggested that total employment in the private security sector was 162,000 but also concluded this would be an underestimate (Jones & Newburn, 1995).

The number and size of forces is not, of course, the only important factor here. We have already noted certain 'upward' or centralising tendencies within policing but it is interesting to note that there are counter-tendencies at work as well. Certainly in the UK and in some of the larger US forces, for example, NYPD, there has been an attempt to decentralise command down to more local areas or precincts (discussed further below in chapters four and five). In part this is a recognition that crime still is primarily a local phenomenon and in part it is a manifestation of a performance culture in which local commanders are required to demonstrate and justify police productivity. The introduction of Compstat in New York City is the clearest example of this (Silverman, 1999, 97-124) but it is an innovation that some UK forces, for example Merseyside, wish to adopt. It is, however, also the case that the rhetoric of de-centralisation exceeds the reality (e.g. Bayley, 1994, 88-91; Loveday, 1999, 359-62).

Networks and policing

There are three different ways in which the production and supply of public and private goods and services might be co-ordinated: by markets, hierarchies or networks, the point being not that one or other of these alone necessarily describes any particular situation but that their use as abstract models and their interaction enhances any analysis (Thompson *et al*, 1991, 15-17). The relevance of these to the issue of crime is examined in the next chapter; here we consider the implications of this for policing. Whether we examine policing as a 'service' or security as an increasingly commodified 'product' (Spitzer, 1987), it can be seen that the balance between these three in determining what citizens/consumers got has varied over time. For example, prior to the nineteenth century provision was largely private and market-driven. The establishment of the 'new police' saw the introduction of hierarchically-organised state police that established a virtual monopoly in the policing of public space although private policing remained central within companies and other institutions. In recent decades, public police have retreated somewhat under the pressure of fiscal constraints and the private sector has re-emerged in the public sphere (e.g. Johnston, 1992). The privatisation of some areas of what was previously viewed as 'public' policing can be seen as a shift from hierarchy to markets but hierarchical organisations play a significant role within the private sector.

In examining certain aspects of contemporary policing and especially currently-developing intelligence structures, however, networks appear to be the most significant analytical tool. There are a number of reasons for this. First, as we saw in chapter one, ideas of governance and steering displace sovereign hierarchical models of government though the significance of the state's periodic use of sovereign coercive power remains an important part of its symbolic claim to authority. With reference to the

UK, Rhodes has argued that changes in British government of the last twenty years can be understood best by defining governance 'as self-organizing, interorganizational networks' over which the centre has only 'loose leverage'. The significance of network analysis is that it stresses the continuity in structural (rather than just interpersonal) relations between interest groups and government departments (1997, 29-59).

Compared with the widespread use of the network concept to discuss European liberal democracies in addition to the UK, Peters argues that there are several barriers to its effective use regarding US government, including the relative persistence of 'iron triangles' that provide a more exclusive structure for government-interest relations (1998, 28-9). However, it is suggested below that the concept will be as much use, if not more, to look at US policing. One obvious objection to the application of the concept to policing (apart from the general criticism that network analysis is simply pluralism in new clothes) is that police are not a pressure group. However, while it may be true that political science has concentrated more on groups as influencing policy-making rather than their role in implementation, and that work on policy networks has continued the imbalance (Kickert & Koppenjan, 1997, 35) this is not a *necessary* feature of the concept. Therefore to the extent that police do seek to influence policy (Savage *et al*, 1999) and are certainly involved, in the broadest sense, in implementing state policy there is no reason not to use the concept.

Marsh and Rhodes have suggested a continuum of types of policy network from the policy community (limited number, some groups consciously excluded, frequent interaction between members, shared consensus on ideology, values and broad policy preferences, all members have resources so their relations are exchange relationships) and the issue network (many participants, fluctuating access and interaction, limited consensus, interaction based on consultation rather than bargaining, unequal power relationships in which many participants have few resources (Rhodes, 1997, 43-5). McCleay used this typology in her study of policing in both Britain and New Zealand and concluded that the general absence of non-state actors in developing policing policy in both countries meant that the policing network was relatively narrow:

> The policing sector exhibits the characteristics of a policy network that is dominated by a core of state employees: policy makers, regulators and practitioners. It is not, however, a policy community. (McCleay, 1998, 130)

Maybe, but it is certainly much closer to a policy community than to an issue network.

The application of the idea of networks (compared with the alternatives of hierarchies and markets) to policing in general, and police intelligence in particular, is not entirely new. The hierarchies within which policing is organised (and the administrative orders that represent their central co-ordinating mechanism) have traditionally provided a formal mechanistic model of police organisation but research has long since demonstrated the inadequacy of the model as a description of what happens. Most of the sociology of policing is built around the detailed observation of the limited extent to which formal policy is actually implemented by the rank and file officers whose discretion is maximised by the structures of policework - its low visibility, the permissiveness of law, especially relating to 'public order' and need of groups of colleagues to protect themselves from the possibility of discipline by supervisors (e.g. Bradley *et al*, 1986). Thus, the actual outcomes of policing have always been at least partly understood in terms of the culture and actions of police 'networking' with colleagues rather than as members of disciplined hierarchies. Therefore, the analysis of networks should incorporate both the formal and informal structures of policing; just as the police themselves make use of social network analysis in mapping the organisation of crime, as discussed in chapter nine.

Even at the formal level the hierarchical model of policing has recently been attacked by the application to it of market models as part of what is called New Public Management (Leishman *et al*, 1996) and there are now crucial areas in which police information processes are market-driven, for example, the purchase of forensic services, but neither hierarchy nor market illuminate intelligence processes as much as networks. First, this is because:

> Networks are particularly apt for circumstances in which there is a need for efficient, reliable information. The most useful information is rarely that which flows down the formal chain of command in an organisation, or that which can be inferred from shifting price signals. Rather it is that which is obtained from someone whom you have dealt with in the past and found to be reliable. (Powell, 1991, 272)

Reciprocity and trust are the name of the intelligence game; information is a 'currency' which one might spend in order to get more. Power thus comes to be defined not in terms of hierarchical office but in one's reputation for having 'knowledge'.

Second, specialist intelligence units are normally small and rely crucially on their ability to trade with other, similar units. Thus, the most effective CIU will be that which networks on all three dimensions of policing: across to equivalent units elsewhere, 'up' or 'down' to another level and across to different sectors (see Figure 2.1 below p.39). Some of these contacts will be legitimated *via* official agreements while others will be officially frowned on, for example, passing information from the Police National Computer (PNC) to an investigator in the private sector, but the efficiency or otherwise with which information flows through these dimensions will be determined much more by the development of reciprocal relations between practitioners than by the presence or absence of formal agreements. Indeed, information can flow through and around a range of formal, legal barriers.

Certainly, formal agreements establishing networks of law enforcement agencies and the circumstances under which they will co-operate both in terms of information-sharing and enforcement activity are an important and growing development at all levels of policing. These usually relate to the initiation of new policing systems and structures that are intended to have some degree of permanency and therefore require specific rules relating to authorisation, financing and operation. Negotiating such agreements can be a lengthy process whether they are transnational, such as Europol, and require the formal approval of national parliaments or local such as the UK Crime and Disorder Act 1998 that requires the collaboration of 'responsible authorities' (local council, police and other agencies) in determining local crime reduction strategies and sharing information accordingly (Home Office, 1998, Ch.5).

But the very formality and systematic nature of these agreements leads practitioners to prefer their own more informal and dynamic networks that respond to their more pragmatic and case-propelled needs (Den Boer, 1999, 71). These may transcend any particular spatial or sectoral divide and typically develop through personal contacts made initially in joint investigations, secondments, people moving to different jobs (within or across sectors), or contacts at conferences. These may involve the exchange of information (that might not be exchangeable formally because of legal protections) and the sharing of gadgetry (to overcome resource constraints) in what has been called 'grey policing' (Hoogenboom, 1991, 18).

Lee argues that these developments are especially significant in the UK because of the pressure on investigative resources and the new emphasis on intelligence-led policing (1995, 389-90). But, based on this study, there is no reason to suppose it is any less significant in North America. For example, Ericson and Haggerty found such networks well-developed in Canada between police commercial units and financial institutions (1997, 203-04). The negotiation of formal agreements such as those referred to above involve, quite rightly, mechanisms for accountability (however weak) that normally involve an audit trail of authorisations. This can lead to a very time-consuming exercise for all concerned as backs are covered. The beauty of informal networks from the point of view of those involved is that this is not required - it is replaced, by trust in the other person's ability to use what you give them in ways that will not come back and bite you. So, as long as it is just an exchange of information that need not become public then it may proceed whether it is formally authorised or not. This can be of use, for example, in 'case-screening', that is deciding which crime reports will be followed up by further investigation (e.g. Lee, 1995, 389).

To the extent that long-standing reciprocal relationships develop, then we can imagine law enforcement intelligence networks in which the closeness of the practitioners to each other may be greater than to the precise policy objectives and interests of the organisations or states to which they formally belong (c.f. Marx, 1987, 187). Yet points of contact in a network may demonstrate conflict as well as agreement (Powell, 1991, 173). Intelligence networks do not necessarily develop as 'seamless webs' of global policing. Conflicts may occur for a variety of reasons, often one participant's uncertainty as to what another is going to do with the information provided. What one sees as an operation furthering the organisational interest may be seen by the other as a precipitate operation hindering greater, if more long term pay-offs. As a general rule one might say that, other things being equal, the prospect of conflict rather than agreement becomes proportionately more likely the greater the *distance* between the points of contact; a recent Home Office study of the policing of cross border crime found similar problems in the making of contacts (Porter, 1996, 22). Instantaneous global transmission of data may be a highly significant aspect of intelligence networks but the facility cannot simply transcend the need for reciprocity and trust.

The three-dimensional expansion of policing networks

Discussion of policing often concentrates on the different 'levels' - local, national, transnational - at which policing structures exist and whether or not they become more 'centralised'.   Loveday, for example, analysed the UK Government's intentions for what became the Police and Magistrates Courts Act 1994 as continuing the centralising tendencies of the Thatcher and Major administrations although they were obliged to water down some of its provision (Loveday, 1996, 22-39). Jones and Newburn, on the other hand, while acknowledging the potential for that Act to be a centralising measure argue that, on the basis of their research, it has not been so with the chief constables rather than Home Office being the main beneficiaries of any power shifts that have occurred (1997, 38-40). In other respects, forces have sought to decentralise their own operations, with additional authority to determine resource-allocation being devolved to Basic Command Units at area level, though, as we saw above, some of these moves have been more rhetorical than real.

It has also been argued that international policing developments reflect a centralising trend, though observers differ on whether they think this either is or should be happening. For example, Hebenton and Thomas see policing as part of the 'broad shift…towards transnational social practices' that is globalisation and also that within this there are hegemonic tendencies, specifically, the pressures from the US on others to adopt US policing practices as a means of reducing those disparities between regimes that are exploited by transnational criminal organisations (1998, 103-07; also den Boer, 1999, 59). But whereas Hebenton and Thomas acknowledge the role of local and national resistance to these tendencies, others seem less confident. *Statewatch*, for example, suggests that now Europol is operational, together with the FBI it 'is poised to dominate the global law enforcement community in the coming decade' (1998, 8(5) Sept-Oct, 21). For others, such as the National Strategy Information Center, this cannot come quickly enough. Its Director argued that US counter-espionage and, later, counter-terrorism strategies provided 'roadmaps' for a US, indeed international strategy against transnational organised crime (Godson, 1994, 163-77) and from a 1992 seminar emerged the suggestion that not just a global criminal court but a global police force was necessary to make up for the inability of nation-states to control the 'new world disorder' (Holden-Rhodes & Lupsha, 1995, 24-5).

It is futile to argue whether one or other of these tendencies is dominant because the most significant aspect of globalisation is that these processes have been occurring simultaneously; a study of health, housing and education in the UK found the same dualism (Pollitt *et al*, 1998, 162-3). This can be characterised as a decentering of policing (cf. deSousa Santos, 1992, 133) that has three main dimensions. The fact that there are now more levels at which policing is organised - from the local area to the transnational - means that we can talk, first, of the 'deepening' of policing: '...developments at even the most local level can have global ramifications and vice versa'. Second, policing is 'stretching': the process by which 'decisions and actions in one part of the world can come to have world-wide effects' (McGrew, 1992a, 3). Most crime and consequently most policing remains local in origin and impact but the significance of markets in, for example, drugs, counterfeiting and smuggling make clear that policing is no more immune from this aspect of globalisation than the political economy in general (cf. Hobbs, 1998, 418-9).

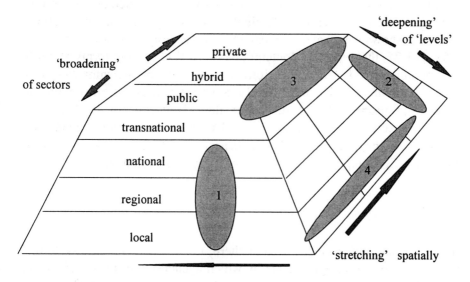

Figure 2.1 Globalisation of Policing Networks

A third dimension is 'broadening' which includes 'the enormously diverse range of agencies or groups involved in...decision-making processes at all levels from the local to the global' (McGrew, 1992a, 3). Whether we describe this diversity in terms of public, private and hybrid sectors

(Johnston, 1992, chs.3-6) or retain the basic public-private divide (Jones & Newburn, 1998, 203-08) there is no doubt that these sectors are increasingly interwoven through joint investigations, hiring or sub-contracting to each other, the creation of new organisational forms and the circulation of personnel (Marx, 1987, esp. 172-3).

Thus, we can develop a three-dimensional model of contemporary policing within which networks develop. This is visualised in Figure 2.1. Network 1 represents public police agencies: these might be more integrated within a national force operating locally, for example, the RCMP in Western Canada), or less so as in the fully-fledged federal policing structure of the US. The relationship between NCIS, the pre-1999 regional crime squads and county forces in the UK could also be represented by this ellipse. Number 2 could be illustrated by one of the large private security companies or the in-house security department of a transnational corporation. Number 3 represents transnational co-operation between police and security organisations in all sectors. There are few if any formal agreements here, but globalisation on the one hand and the movement of personnel between sectors on the other certainly facilitates increasing informal interactions. Number 4 shows what have been called local 'security networks' (Johnston, 2000, 163-75) that represent contemporary efforts at varying forms of multi-agency policing.   These may be formal, as with the partnerships envisaged in the UK Crime and Disorder Act 1998, or more informal 'grey policing' networks of information and 'sub-contracting' (Hoogenboom, 1991, 17-30).

Aspects of developing networks in the UK and Canada are discussed in chapters four and five respectively; here New York State is taken as a US example and illustration of the general argument.

Formal networks in New York

In the US law enforcement system there are many more or less formal information-exchange agreements by which agencies seek to overcome not only its fragmentation but also the lack of resources available to the majority of smaller agencies. Given the New York State Police's (NYSP) position as state-wide it provides a *convenient* perspective on the major agreements in place within New York but this should not be taken as an empirical statement that it is the central locus of these agreements.   For the large number of small up-state agencies NYSP does provide an important point of communication with other local, state and federal agencies but, on many occasions, other agencies will by-pass it and deal with each other either formally through one of these agreements or informally.   Figure 2.2 presents a view of the most significant formal information- and intelligence-sharing agreements that have been established with a view to facilitating exchanges both horizontally between agencies at the same 'level' and vertically with those at other (local, state, federal, transnational) 'levels'.

NYSP was established in 1917 and with 4,100 sworn officers (troopers) is one of the ten largest forces in the US. The Crime Analysis Unit is located within the Bureau of Criminal Investigation and has six sworn and three civilian analysts. Much of their work relates to narcotics and is reactive to requests from people in the field. Many of these will be from the 300 officers working in narcotics and the Special Investigations Unit (for organised crime). Officers have state-wide jurisdiction but priorities in the field will be set differently depending in part on the area concerned. NYSP is in some respects similar to the RCMP: in some, predominantly rural, areas it is *the* police, where other agencies also operate, priorities will be set in part by a process of negotiation with them.

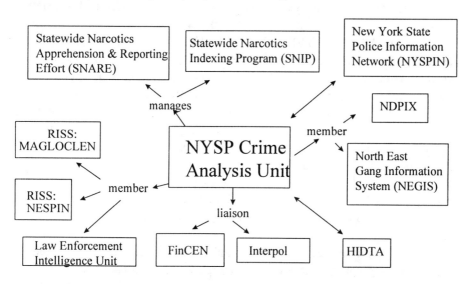

Figure 2.2 New York State Information and Intelligence Network

The precise nature of the information agreements covered in Figure 2.2 and the NYSP Crime Analysis Unit's relationship to them varies. There are, broadly, two main types: databases and pointer indexes. The latter are essentially an attempt to prevent duplication of effort by providing agencies with a means of finding out whether some other agency has an ongoing investigation or information regarding a particular target. As such they facilitate bilateral information swaps between officers. NDPIX was developed by the federal DEA and requires participating agencies to submit details of their active cases (name, sex and age or date of birth) in return for 'pointer' information to agencies elsewhere with the same targets. The system is not a database since case information can only be exchanged by people once put in touch with each other. If a positive match is found and sent, the originator of the record is also notified. The Statewide Narcotics Indexing Program (SNIP) is a similar state-wide pointer index, managed by the NYSP, that is seen as more robust than NDPIX since it can be used with fewer known data fields (deBlock, 1999).

The New York State Police Information Network (NYSPIN) is essentially a message-switching system managed by the State Police for the transmission of alerts, criminal history information, and access to records regarding, for example, vehicles and wanted persons. It also interfaces with other state and national systems such as the National Crime Information Center and databases maintained by the Division of Criminal Justice Services in Albany (NYSP, 1997, 59). The Division is responsible for collating the Uniform Crime Reports for New York and also runs a computerised criminal history and fingerprint system that theoretically receives all arrest and disposition information from agencies (Rosen, 1999). This also facilitates a narcotics database (SNARE) managed by the State Police into which all police agencies throughout the state are encouraged to submit information relating to drug seizures and arrests. One of the products of the analysis of the information put into the system is a quarterly newsletter of about 16 pages containing information concerning, for example, legal developments, concealment techniques, summaries of data on the origins and destinations of intercepted drugs and prevailing street prices.

The State Police provides the link for small agencies to a number of cross-state networks. The earliest attempt to systematise intelligence co-operation between law enforcement agencies came in 1956 with the setting-up of the Law Enforcement Intelligence Unit (LEIU) by James Hamilton of the Los Angeles Police Department Criminal Intelligence Unit. The objective of the Unit was:

to promote the gathering, recording and investigating and exchange of confidential information not available through regular police channels, concerning organized crime. (quoted in Donner, 1990, 80)

A specific frustration with the 'regular channels' that prompted this initiative was the resentment among local agencies that the FBI would not grant access to their files but would happily make use of their information if they could get it. Accordingly, in pursuit of its anti-federal philosophy, federal agencies have not been permitted to join as regular members.

LEIU rapidly accrued over 200 members nation-wide but still has fewer than 300 members and is administered from within the California Department of Justice. Yet, despite this and the fact that its member agencies are public bodies, LEIU has always been a private organisation; this has enabled it to avoid some of the restrictions placed on public police in terms of intelligence files. The main function was to act as a clearing house for intelligence 'face cards' that included photos, date of birth, description, 'hangouts', *modus operandi*, arrests and associates. Although established specifically with organised crime in mind, through the 1950s and 1960s LEIU's attention shifted towards counter-subversion and public disorder by 1970.

LEIU's role changed somewhat in the 1970s: although remaining private, it received Justice Department grants in 1971 and 1974 to assist the computerisation of its Interstate Organized Crime Index and the dissemination of other data. Receiving public funding led to some limitation on its ability to process political intelligence, though the manual system remained unaffected. Further, along with the FBI and a number of city police intelligence units, LEIU found its practices coming under the scrutiny of the courts (Donner, 1990, 80-4) and in 1974 the Justice Department initiated the Regional Information Sharing Systems (RISS), of which there are now six, and which now dwarf LEIU in terms of resources and membership. Now, LEIU provides a pointer index and training for its members and proselytises the cause of intelligence within the law enforcement community.

The RISS now have over 4,500 law enforcement agencies as members (that is, about 25% of the total). The main objectives are:

to encourage and facilitate the rapid exchange and sharing of information pertaining to known or suspected criminals or criminal activity among federal, state and local law enforcement agencies, and the enhance co-ordination and communication among those

> agencies in pursuit of criminal conspiracies determined to be interjurisdictional in nature. (Gallagher, 1996, 23)

Information sharing and the development of analytical capabilities are the main priorities and in 1996 there were 46 analysts assigned to the six regions who sought to provide analytical services and training for members. The second generation of RISS technology was to become available in 1996 and would provide a national network - through connecting the regional systems - incorporating a database of 475,000 subjects with free text inquiry, though initially members' inquiries would have to be made by telephone. Until 1996 project analysts were able only to respond to specific requests from members; as the ICT progresses and member agencies themselves up-grade it is clearly hoped that analysts will be able to carry out more thorough and national analysis (Gallagher, 1996). The State Police is a member of two of the regional projects: MAGLOCLEN with 440 member agencies in the mid-Atlantic states and including Ontario and Quebec and NESPIN that includes 360 agencies in New England. These regional projects are also involved in public-private information-sharing projects.

Another public network that has evolved in the last few years is the North East Gang Information System (NEGIS) that includes New York and all the New England states except Maine. The respective state police forces are the access points to the system which seeks to be, first, a secure mechanism for the transfer of information between member agencies. Local agencies have to sign a Memorandum of Understanding and obtain the appropriate hardware and software and there is a training requirement. Second, however, the information traded is not restricted to gangs - there is no real restriction on what can be shared and it is also building a database.

The Anti Drug Abuse Act 1988 that created the Office of National Drug Control Policy to be headed by the so-called drug 'tsar' is another federal attempt to increase co-operation in which the federal carrot is money. One initiative has established High Intensity Drug Trafficking Areas (HIDTAs) in various regions since 1990 to assess drug threats, design counter strategies and develop implementation initiatives. HIDTAs thus seek to co-ordinate local, state and federal programmes by means of an executive committee that seeks to balance local/state with federal represent-atives. There are now 21 designated regions of which New York/New Jersey was one of the first (www.whitehousedrugpolicy.gov/). Funding is concentrated on those agencies and projects that show willingness to co-operate and share information. For example, NYSP is receiving money from the New York/New Jersey HIDTA to develop a photo-imaging system to permit the computerised dissemination of 'mug-shots' (deBlock, 1999).

Moving to the national and transnational level, the State Police has direct relations with federal law enforcement agencies such as the FBI, DEA, Customs and INS that can involve information sharing and, in some cases joint operations. (See chapter five for discussion of the main federal law enforcement agencies.) But it also acts as liaison with other information networks, the two identified as examples in Figure 2.2 are FinCEN and Interpol. The Department of the Treasury operates the Financial Crimes Enforcement Network that was established in 1990, has some 200 employees and seeks to act as a bridge between police, other regulatory agencies and the banking sector in the control of money-laundering. FinCEN has developed its own database and has made extensive links with others in the business and banking sector in order to fulfil its main role of information gathering. It does have investigative powers but is more concerned to assist agencies with their own investigations. The basic source of its data is the compulsory reporting of all transactions above $10k that is required of financial institutions. By 1995 these amounted to 11.5 million per year (Thony, 1996, 259, 269). A recent Government Accounting Office study of FinCEN reported that law enforcement agencies value both the tactical support and financial information that they are able to obtain (GAO, 1998).

The State Police also provides a liaison point for state agencies with Interpol. Not considered to be much more than a rather insecure post-box for law enforcement information until ten years ago (Hebenton & Thomas, 1995, 64-9), in the face of the development of Europol, Interpol has struggled to improve its image and to some extent seems to have succeeded (e.g. Hoey & Topping, 1998, 531-2). It has sought to shrug off the post-box image *via* the development since 1993 of an Analytical Criminal Intelligence Unit (Interpol, 1998, 48) and is currently encouraging the development of regional networks for information sharing, initially in Europe and South America, that it hopes will be more secure (Gilligan, 1999). US involvement in the development of Interpol was patchy, mainly because of Hoover's hostility to the organisation during his long Directorship of the FBI - Hoover was notoriously unwilling to collaborate with any other organisation over which he did not have control. Consequently the Treasury department was the most significant US participant until the 1970s. Then, in 1976, as the FBI re-positioned itself following the traumas of congressional investigations, a bureaucratic 'bloodbath' ensued between Treasury and Justice as to where the National Central Bureau should be located. Justice 'won' but since then an uneasy truce has prevailed with control of the national office shifting between nominees of the two departments (Bresler, 1993, 105-18; see also Nadelmann, 1993, 181-6).   The additional liaison offices within each state police force have been developed since the 1980s (Bresler, 1993, 208).

Informal networks in New York

There are three important dimensions of information exchange that do not appear in Figure 2.2.   First, a great deal takes place informally between officers who know and trust each other - this is crucial since once information has been passed to another agency the originating agency has lost control of what happens to it. One interviewee described how he and several of his staff did extensive lecturing about gangs for the regional projects largely because the resulting networks of contacts developed were so useful (Arsenault, 1999). All networks depend fundamentally on trust: this is why more information will always be traded bi-laterally compared to the amount that will be put into remote databases in which the ultimate recipient is unknown. Pointer indexes seek to formalise this by providing a mechanism for individuals to be put in touch with each other.   The further co-operation is to be taken, the more important personal relationships become: if information-sharing is to move on to a joint operation (requiring the commitment of greater resources and greater transparency) then the relationships between the different commanders become crucial (deBlock, 1999).

A second dimension results from the formation of multi-agency taskforces targeted on a specific problem. This way of seeking to overcome organisational fragmentation is not new - at the federal level they were a feature of Attorney General Robert Kennedy's campaign against organised crime in the early 1960s (Goldfarb, 1995, 310-11) and they continue in the HIDTAs (see above).   Similar structures can be seen in the UK, for example the Terrorist Finance Unit in Northern Ireland and the Joint Action Group on Organised Crime set up by the Metropolitan Police in 1992 (Norman, 1998).   An important aspect of them is that they enable law enforcement to overcome barriers to information flow because officers on assignment will retain their access to their home agency's database and thus the agencies' databases are effectively linked, at least as far as the specific targets of the taskforce are concerned. Clearly this may well increase the operational potential of the agencies but it does subvert any restrictions placed on information transfer for reasons of privacy. In the longer term, to the extent that taskforces develop their own agendas and priorities, then arguably they may be seen to reinforce fragmentation rather than reduce it by linking agencies.

Establishing a taskforce requires a high degree of formality including memoranda of understanding agreed to by the contributing agencies. The New York State Organized Crime Taskforce (NYSOCTF) is one such and is plugged into the formal networks shown in Figure 2.2. But it has additional systems, for example, an 'informal' pointer system concerning New York City that is run by the DEA. A minor but recurring theme of interviews with local and state officials was the hesitancy in dealing with federal agencies. As we saw above, the state pointer index was preferred in some ways to the DEA-managed NDPIX and there was a suggestion that local and state agencies fear that DEA use pointer systems to poach good information or cases. Similarly, the difficulty of obtaining routine access to federal databases meant there was a reluctance to pass information up to them. A similar effect can be gained more quickly and informally by the simple expedient of officers being posted to other agencies - numerous examples of this emerged in the current study.

Informal information exchanges are so extensive because practitioners find them indispensable but, precisely for this reason, they raise issues of accountability and privacy. The local and national controversies in the early 1970s in the US regarding political intelligence gathering by police did lead to greater restrictions on the circumstances under which information could be gathered, stored and disseminated, for example, in New York in 1973, Los Angeles in 1975 (Donner, 1990, 191, 266-8) and for the FBI in 1976 (Elliff, 1979). Such guidelines can also be incorporated into a Memorandum of Understanding (MOU) that governs formal co-operation and information-sharing between agencies but the extent to which they actually govern police practices may be less than their role in providing *post hoc* presentations of those practices (see Dixon, 1997, 299-311 for general discussion on the relation between rules and policing).

Other things being equal, this is more likely to be true, the greater the informality of information-exchanges. For example, state law in New York provides for greater restrictions on the exchange of arrest records than most states. If, after arrest and fingerprinting, a person is acquitted or convicted only of a lesser traffic or local ordinance violation, the court must notify the law enforcement and prosecuting agencies involved in the case that the records must be 'sealed'. This means that the defendant's fingerprints and photograph must be destroyed or returned to her; the object of this process is to return the person to their status prior to arrest or prosecution. The numbers of records that are consequently fully or partially sealed can be large, for example, by July 1998 it was 49% of the 576,000 'arrest events' in 1996 (DCJS, 1998, 3-6). To the extent that arrest records normally form a significant part of police intelligence systems, these restrictions are a barrier to the development of such systems. Potentially, however, this barrier, like others, may be overcome by informal personal contacts. Indeed, the issue of the impact of sealing on intelligence processes has not really been considered.

Transnational policing networks

The most all-encompassing factor within which current developments in policing are discussed is globalisation and the effect of this is represented in Figure 2.1 as part of the three-dimensional extension of policing networks. Inter-related with globalisation but existing as a distinct process has been the 'regionalisation' of states with an intensification of their economic and political interactions, for example, the North American Free Trade Area (NAFTA) and the European Union (EU). These agreements are the most developed of a range of multilateral associations, reflecting a 'desire by most states for some form of international governance and regulation to deal with collective policy problems' (Held & McGrew, 1993, 271).

Prior to the 1990s most discussion of transnational intelligence structures revolved around the security networks erected as part of the Cold War and, in the West, under US hegemony (Whitaker, 1998). Otherwise transnational policing co-operation was relatively underdeveloped - since the 1920s it had been manifest mainly *via* Interpol and did not develop any more significant role hampered at it was by the lack of trust which is necessary for any meaningful intelligence network to develop.

Within NAFTA there is constant US pressure on its partners to improve security. For example, Canada is alleged to be particularly lax at its borders resulting in much illegal immigration and infiltration by 'terrorists' seeking an easy route into the US. Most recently, in December 1999, an Algerian who had been living in Montreal was arrested at the US border with explosives in his car. Ignoring the obvious point that his appre-hension and the lack of the predicted millenium terrorist campaign in the US suggested that border security was not a major problem, a combination of budgetary and election politics sparked off a particularly virulent debate regarding Canada's alleged laxity. The full imposition of the 1996 US Immigration Reform Act scheduled for March 2001 is already intended to impose greater entry and exit checks and, although its impact remains to be seen, the creation of 'Fortress North America' to parallel that in Europe is clearly high on some political agendas. The sheer length of the US-Canada border (7,000kms) is part of the problem and both US Customs and Border Patrol announced increases in the numbers of agents they have there - 1200 and 300 respectively - in the wake of the latest row (*National Post*, September 18, 1999; January 27, 2000; *Ottawa Citizen*, January 27, 2000).

But there can be few finer examples of the contradiction inherent in states' efforts to liberalise licit markets while prohibiting illicit ones than the current situation on the US Southern border with Mexico where there are many more agents: 2000 Customs and 7,400 Border Patrol. Until the mid-1980s Mexico represented a stable and effective, if not especially democratic, regime, organised around the *Partido Revolucionario Institucional* (PRI) and consisting of a complex of familial networks, political clientelism, populism and orderly corruption (Castells, 1997, 277-86 on which this summary is based). Within ten years, however, what had seemed so stable for sixty years, collapsed. First, a debt crisis started in 1982 and the resulting austerity programme induced a recession that alienated organised labour and other urban groups who had been major props for the regime. The state was facing two main pressures to liberalise: first from various international institutions such as the International Monetary Fund in return for financial assistance and second, from the US in the context of preparations for the establishment of NAFTA in 1994. Political support for the PRI reduced further following the protests at the official response to the 1985 Mexico City earthquake and the narrow victory of their Presidential candidate in 1988 was more obviously due to fraud than usual. The new President, Carlos Salinas, responded with further cuts in public spending, widespread privatisation and opening the country to foreign investment. The last of these did increase rapidly while the standard of living plummeted for most people. The signing of NAFTA in 1993 symbolised the integration of Mexico into the global economy.

Immediately the country almost collapsed - the Zapatista insurgency in Chiapas began on January 1, 1994 and exposed the reality of the exclusion of the peasants and Indians from the PRI regime. Within the year the currency collapsed and Mexico almost defaulted on its debt, the money laundering activities of Raul Salinas, the President's brother, were exposed and he was jailed for his part in the murder of a senior PRI official and the party's next presidential candidate was also murdered. Castells concludes that both these murders were orchestrated by the developing alliance of parts of the traditional PRI apparatus and the drug traffickers that played a decisive role in the 1994 crisis, thus illustrating how 'the globalization of crime overwhelms powerful, stable nation-states' (1997, 284).

Given the violence of the denouement in 1994 this is an understandable conclusion but distinguishing cause and effect is difficult, especially in this case because the processes of liberalisation and the increasing significance of the drugs trade in Mexico coincided during the preceding decade. In part because of the growing US enforcement pressures against Colombian traffickers in the Caribbean, there was a rapid increase in the significance of Mexico as the route for drugs importation into the US; it is estimated that the proportion of Colombian cocaine trans-shipped through Mexico increased from about 20% to 70% between the mid-1980s and mid-1990s. But there were other unintended though foreseeable consequences of liberalisation that facilitated drugs trafficking: the de-regulation of the trucking industry in Mexico in 1989 and the NAFTA requirement that Mexican trucks will eventually be able to travel throughout US and Canada; the post-1988 privatisation of state assets provided a magnet for investors including those needing to launder the proceeds of drug sales and the reduction of agricultural subsidies increased significantly the number of peasant farmers for whom growing marijuana and opium was the best or only alternative (Andreas, 1999, 133-6).

Thus, at the same time that Mexico sought the creation of a minimalist state regarding the legal economy, a maximalist state was being created to regulate the illegal drug economy. As intended, NAFTA led to a rapid increase in the trade between the two countries: the value of imports from Mexico to the US doubled between 1993 and 1997 but that included significant increases in both illicit and licit trade though quantifying the proportions of each is no easier for analysts than separating them is for law enforcers. Since 1988, successive Mexican Presidents have declared drugs trafficking to be a national security threat and have reinforced the military's antidrug role as part of an expanding national security apparatus (Andreas, 1999, 125-36) that is directed also at labour migration to the US. This dates back to the nineteenth century but has increased since the 1960s as the industrialisation of the Mexican side of the border acts as a magnet for labour that, now further reinforced by NAFTA, may then migrate, often only temporarily, into the higher wage US (Andreas, 1996, 59-62). This was not the intention since, unlike the EU, NAFTA prohibits free movement of labour; indeed, one of the arguments used to sell the agreement in the US was that it would reduce Mexican migration by improving the Mexican economy (Whitaker, 1998, 430).

The resulting effort to secure the border is a complex network of agencies that seek to deploy increasing numbers of personnel. For example, INS personnel increased by almost one-third between 1993 and 1996 (see Table 5.1 in chapter 5) and agents in the Border Patrol on the south-west border increased by 65% between 1990 and 1996. In February 1995 Customs announced an increase in 20% in resources devoted to border inspection called 'Operation Hard Line': 'This is a war' said the Customs Commissioner (Andreas, 1996, 62). In 1998 Customs and INS together launched a further Border Co-ordination Initiative aimed both at drugs and illegal aliens (www.customs.ustreas.gov 24/4/99). The US side of the border presents in microcosm the law enforcement situation in the US generally, that is, a highly fragmented network of agencies that may sometimes compete, sometimes co-operate but often may simply be unaware of what the other is doing. The Southwest Regional HIDTA includes the border states (Texas, New Mexico, Arizona and California) and, as in New York and New Jersey (see above) is another federal effort to use the funding carrot to encourage agencies to co-operate. In Arizona, for example, the HIDTA located in Tucson seeks to co-ordinate the investigation of drug smuggling in six border counties. An Intelligence Division is staffed by analysts and investigators from ten different local, state and federal agencies, each making available its agency's data and using its agency's system for the dissemination of information at the request of another participant (Martinez, 1997, 13-15).

The 'war' is to be fought with all available technology. South of San Diego a ten foot high wall has been built from steel plates originally designed to create temporary runways during the Gulf War while various devices such as magnetic footfall detectors and infrared body sensors originally used in the Viet Nam war are deployed in more remote stretches. A Border Research and Technology Center was opened in Southern California in March 1995 to further develop and adapt military technology to the problems of border control (Andreas, 1996, 63). Even the National Guard is involved, having set up its own National Interagency Civil Military Institute aiming to develop training for both law enforcement and military agencies in interagency antidrug operations (www.nici.org 24/4/99).

On the Mexican side policing problems are aggravated by the chronic corruption of agencies in border areas. For example, in Nuevo Laredo, a city whose population has doubled to 400,000 since 1994 a new police chief took office on January 1 1999 and within six weeks had dismissed 130 of the police department's 600 officers in an attempt to rid it of corruption that saw police involved in robberies and acting as body-guards for drugs traffickers (*New York Times*, February 16, 1999, A15). Still, corruption is not exactly unheard of on the US side, notably Los Angeles.

The maximum and minimum states 'meet' each other once a year when the US President has to certify to Congress that Mexico is a reliable partner in the US 'war on drugs'. A negative recommendation would set in motion US trade and economic sanctions against Mexico that would send the country's bi-lateral relations and NAFTA into crisis since Mexico is the US's second largest export market; consequently there is much political pressure for a positive recommendation whatever the evidence (*New York Times*, February 16, 1999, A1). For example, DEA head Tom Constantine described Mexican drug trafficking organisations to US Senators as posing the worst criminal threat that he had seen in his 40 years in law enforcement (*New York Times*, February 25, 1999, A9) yet certification proceeded. The two main beneficiaries from economic liberalisation in Mexico seem to be the criminal networks involved in the trafficking of drugs and the border security complex that seeks to regulate them, the agencies involved have been able to protect and enhance their budgets more successfully than most (cf. Andreas, 1996, 65 and see Table 5.1, p.120).

Within Europe information-sharing networks are more systematic and have been developed since 1976 with, first, the institution of Trevi as an inter-governmental structure (separate from the then European Community institutions) aimed particularly at intelligence regarding political violence and public disorder. In 1985 the Schengen Convention was agreed to by five of the then twelve Community members and, once the UK and Ireland join, will be co-terminous with the EU. What is distinctive about the Convention is that it combines both detailed rules as to the functioning of the computerised Schengen Information System and prescriptive rules regarding asylum seekers and immigration. The information system is an internationally accessible database incorporating six main categories: people wanted for arrest for extradition purposes, aliens reported so they can be refused entry, people whose whereabouts are to be reported, people summoned as witnesses or suspects in criminal proceedings, people to be subject to discreet surveillance and, finally, objects to be seized. As of

March 1999 there were 1,239,055 persons included in the database. Development of the project has been dogged by criticisms of the lack of transparency and accountability - Schengen represents just one part of the 'democratic deficit' in Europe's developing security complex (Hebenton & Thomas, 1995, 59-64, 174-79; *Statewatch*, 1999, 9(3&4), 22-4).

The Maastricht Treaty 1992 for the first time brought aspects of these developments within the third pillar of the EU treaty (justice and home affairs) though there remained a considerable democratic deficit in that the opportunity for elected representatives to contribute was highly restricted. The most formalised attempt to transnationalise policing in Europe came with the introduction of the European Police Office (Europol) which was envisaged by most EU countries as essentially an intelligence office that would, for the first time, provide a central pool of law enforcement intelligence, though Germany, and especially Helmut Kohl, retained the ambition that Europol would become an operational police force throughout Europe. The official view in UK seemed to be that this was extremely unlikely for a number of reasons, including the lack of a common criminal code. Yet at the Amsterdam summit in 1997 an agreement was reached which clearly envisaged a role for Europol in the operations of European police forces (*Statewatch*, 1997, 7(2), 2).

However, in general it is clear that the movement towards some formal co-operative networks of law enforcement have been relatively slow. There are a number of important reasons for this. Whitaker suggests that one reason is the extensive bi-lateral agreements already in existence from the Cold War that have been developed from below by practitioners with minimal involvement of ministers and which, therefore, suit practitioners very well and make the search for grand international agreements less urgent (Whitaker, 1998, 431-2). A long-standing but still developing example of this is the use of liaison officers, the most extensive example of which is Customs' network of Drugs Liaison Officers; they also now have eight Fiscal Liaison Officers to deal with tax and excise matters, mainly in Europe but the most recently-appointed being in Hong Kong.

Even if formal multilateral agreements are seen as, in some respects, preferable, their negotiation will be fraught with difficulties since internal security and law enforcement are precisely those areas of traditional state power where politicians are most wary of being seen to give up their 'sovereign' powers. Bi-lateral arrangements are further augmented by informal networks of exchange that have developed within the law enforcement and security intelligence communities. One organising framework for law enforcement co-operation is a 'Group of 20' - the fifteen EU states plus the US, Canada, Norway, Australia and New Zealand - that is seeking the creation of a global system for the surveillance of telecommunications - phones, faxes and e-mail *Statewatch*, 1997, 7(4&5), 1-2). Another is the International Working Group on Police Undercover Activities that provides a forum for mutual assistance in undercover operations (Penrose, 1996).

There is now an extensive literature regarding many of the issues raised by these transnational developments (Anderson *et al*, 1995; Hebenton & Thomas, 1995; *Statewatch, passim*); for present purposes it is important to note that intelligence processes are absolutely central: networks of information exchange can develop *sub rosa* in a way that joint operations cannot. The former can be conducted quietly, informally with no other requirements than mutual trust between donor and recipient since there need be no public manifestation of the exchange at all. The latter, however, require *action* that, even if conducted covertly, carries with it the much greater risk of becoming public knowledge and therefore something for which an agency might be held to account. Therefore, intelligence networks are always likely to be far in advance of operational networks and informal ones will always will be far more extensive than formal ones.

Conclusion: a 'geology' of policing?

It is suggested that the best way of understanding contemporary developments in the organisation of policing is to view them as networks expanding at local, national and transnational levels and across the public-private divide. The various levels at which policing is organised within the three-dimensional model presented so far can be compared with rock strata of differing permeability in terms of their ability to absorb information and the extent to which they facilitate or inhibit its flow. Where the strata meet, that is, the organisational boundaries between intelligence units in different areas, we find considerable 'leakage' from the policing process. The extent of attrition within the criminal justice process is now well-documented: as we saw in chapter one, in the UK only 27% of offences are recorded by police and 5% cleared up.

Figure 2.3 illustrates this - at the first two levels of the diagram the proportions of crime that are not recorded or investigated can be empirically supported; thereafter the representation shows how the voids in the system exist at a number of levels without claiming that their size is empirically validated. Claims by the police themselves that such voids are being greatly reduced (HMIC, 1997b, 26) are little more than rhetoric - the vast majority of crime never has been, is not now and never will be subject to investigation. In this view, there is no technical solution to overload other than that most information lies dormant (other, possibly, than for the production of aggregate data) until it is specifically targeted as being of interest.

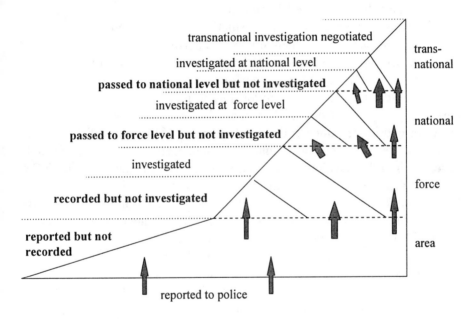

Figure 2.3 Attrition within the 'geology' of policing

A key tenet of intelligence-led policing is that police should not waste resources on reacting to reports of those crimes that not solvable on the grounds of a lack of forensic or identification evidence. Rather, police should target 'known' criminals who are believed to be responsible for a disproportionate amount of crime. Accordingly, Crime Management Units now routinely 'screen-out' from further investigation at least half of the crime reports they receive (Amey *et al*, 1996, 22-3; Williams, 1997) by dealing with victims over the telephone. From now on 'up' through the levels of policing there are a series of further screens through which crime passes. In Figure 2.3 the emboldened headings each represent points at which crime reports leak away from the system of active investigation. At each level cases may be dealt with, passed on up or be left in a 'void' of inaction. In North Wales, as we shall see, the local area will deal with reports involving its own local, 'level one nominals' while passing on cases involving 'level two nominals' to the force level. Some proportion of these will be targeted by the FIB and possibly be subject to operational intervention, others will not be dealt with. The size of this void existing between areas and force-level has increased to the extent that power and resources have been devolved down to area level.

Those investigations for which the Force considered it had inadequate resources or which involved inter-force crime prior to 1999 would be sent up to the Regional Crime Squad (RCS). But it was the RCS that decided which cases it would pursue and, in the nature of things, their resources were devoted disproportionately to major urban cases than those emanating generally from smaller, rural forces. There was considerable criticism within the police at the 'ratcheting up' of the work of the RCS as they and regional criminal intelligence officers became increasingly absorbed into the developing national structure represented, first, by NCIS and now also by the National Crime Squad (NCS). Similar concerns have been expressed before, for example, the Police Foundation's study of drugs squads found the same void between force drugs squads and RCS drugs wings (Wright *et al*, 1993, 44). The current criteria for targeting of the national squads are discussed in chapter six. Eventually the concern at the growing gulf between what forces are able to deal with and what the RCS/NCIS nexus targeted led to the establishment by the Association of Chief Police Officers (ACPO) of a study of inter-force crime and a subsequent conference in Manchester in January 1997.

Associated research carried out by the Home Office found, for example, that investigating officers found their superiors reluctant to commit resources to the investigation of someone who offended largely outside their home area and their inter-force crime was frequently not followed up unless a target was flagged by NCIS or deemed worthy of RCS attention (Porter, 1996, 21-3). Cross-border operations *do* take place, operations involving both local and regional/national squads *do* take place and so do joint operations between public police and the private sector, but they represent a small proportion of the potential for a variety of structural and resource reasons.

What this means in terms of our evolving 'geology' of contemporary policing is that policing (in whatever sector) is not just organised in layered strata across many areas and that the divisions that exist are not just boundaries as implied by Figure 2.1; rather, they are spaces or voids into which a large number of 'crimes', 'targets', potential cases or whatever, disappear. If policing is best represented by the idea of a network then it is very far from an integrated one; it resembles much more the dis-integrated network shown in Figure 2.3. From time to time 'bridges' such as interagency task forces or 'project groups' are constructed across the voids but they may only be temporary.

Developing policing networks reflect the intentional actions of practitioners to overcome the disintegrated structures of police organisations (in all sectors). Networks can be seen as far more useful for mapping these developments than either hierarchies or markets although the significance of the latter have increased in line with the fiscal crisis of the state. But whether network analysis can do more than map intelligence-led policing remains uncertain. We must beware of over-endowing networks with significance, for example, confusing the impact of the networks with what might be simply the actions of its constituent members acting either in conflict or in co-operation (cf. Dowding, 1995, 137; Peters, 1998, 24-5). In chapters four and five we develop some further 'maps' of the networks in, respectively, UK and North America and then go on to examine the practices and processes of networking that give the structures of law enforcement intelligence their dynamism (Hay, 1998, 34-5). In the next chapter, however, we discuss some important issues relating to the organisation of crime and the law enforcement community's perception thereof.

3 Crime Networks

Introduction

The law enforcement networks discussed in the previous chapter may potentially deploy intelligence techniques against any kind of illicit activity, from the minor to the major. However, the growth in the significance of intelligence-led policing has coincided with increased debate (and some might say, panic) about organised crime. This concept has been used to describe the more serious criminal phenomena in the US at least since the 1920s, in the UK it has a more recent history. The term itself has not been used (and therefore required no definition) in the relevant UK legislation but, at the operational level, 'serious and organised crime' is now a much-used phrase. The other significant development since the official end of the Cold War has been the rapid emergence of 'transnational organised crime' as a phenomenon that is presented as a threat to the very security of states. This chapter examines the major parameters of the debate concerning the definition of organised crime with particular reference to the issue of the interaction between law enforcement and those activities it seeks to control. It is important to consider the complex nature of the organisation of illegal activities in order to challenge the fragile basis for panics generated around the activities of criminal 'godfathers' and, though this can seem overly semantic to practitioners, it is important for them also because how the 'problem' is defined and measured determines what, if anything, can be done about it (cf. Levi, 1998).

Definitions of organised crime

There are two general factors included to a greater or lesser extent in definitions: particular *substantive* criminal offences that seem most amenable to being committed in an organised rather than disorganised fashion or the *process* of committing crimes. Some official definitions combine elements of both, for example, the US Omnibus Crime Control and Safe Streets Act 1968 said:

> Organized crime means the unlawful activities of members of a highly organized, disciplined association engaged in supplying illegal goods and services, including but not limited to gambling, prostitution, loan sharking, narcotics, labor racketeering, and other unlawful activities of members of such associations. (Passas, 1995, xv)

Two years later, Congress enacted the Racketeer Influenced and Corrupt Organizations Act (RICO) that aimed to empower prosecutors by liberating them from the need to prove individual culpability for discrete criminal acts or conspiracy. It proved highly controversial in application and has been used in a wide variety of cases to target multi-defendant groups involved in some pattern of criminal activity. The two key elements of this are the presence of an 'enterprise' (any individual, partnership, corporation, association or other legal entity, and any union or group of individuals associated in fact although not a legal entity) and a 'pattern of racketeering activity' that requires at least two acts of racketeering to have been committed within the last ten years, one of them within five years of the indictment. By 1995 there were 45 specific crimes identified as 'predicates' for a prosecution including murder, arson, bribery, extortion, fraud, obstruction of justice and money laundering (Ryan, 1995, 85-9, 173-6).

'Organized crime groups' may be distinguished from 'gangs of ordinary criminals' in terms of the offences they typically commit: the latter, for example, commit burglary and armed robbery that are 'predatory' in that they involve the redistribution of already existing wealth while the former commit more 'enterprise' crimes involving the production and distribution of new goods and services and thus involving more complex market relations (Naylor, 1997, 3). Of course, this division is not always clear-cut, and, the more opportunistic criminal groups are, the less it may be so, but the distinction remains a useful one.

Some analysts have developed this distinction into developmental models of organised crime. Peter Lupsha identifies three stages through which criminal groups may move: first, the 'predatory' in which they are street gangs whose criminal activities are directed at immediate rewards and are relatively vulnerable to law enforcement. Second, a 'parasitical' phase in which corrupt interaction with the overworld increases, for example, as Italian-American crime did during Prohibition. The third phase is 'symbiotic' in which the bonds between the criminal group and the political system become mutual as, for example, in the criminal control of the Fulton fish market, construction and garbage industries in New York City (1996, 30-32). Paddy Rawlinson develops an essentially similar model regarding organised crime in Russia (1997, 29-32) but Tom Naylor argues that the American and Russian experiences were actually sharply different. In Russia, he suggests, Lupsha's model was reversed: the main role of organised criminality was to grease the machinery of the central plan and there was little space for enterprise crime. The parasitical stage developed from the 1960s onwards, largely as a provider of consumer goods that the formal economy could not and by the 1980s the intensifying collapse of the central system fuelled by an explosion of drug trafficking following the Afghan war, a prohibition-like anti-vodka campaign and eventual privatisation culminated in a combined assault on state authority from criminals and state officials (Naylor, 1995, 46-49).

There is certainly a danger of imposing neat developmental models upon a phenomenon that is, in its essence, pragmatic, fluid and opportunistic. Police and prosecutors are most likely to define organised crime in terms of specific offences to the extent that the intended outcome of law enforcement is prosecution. But prosecution becomes less important and disrupting criminal activity increases in importance as an objective of law enforcement (see further in chapter nine). In the UK the criminal law concerns itself almost entirely with substantive offences, the main exceptions being incitement, conspiracy and the Prevention of Terrorism Act. There is no UK equivalent of the RICO law directed at the suppression of criminal enterprises *per se*. But even if substantial offences are the main concern, a second way in which organised crime is defined is in terms of certain defining characteristics, for example, Interpol have developed a list of characteristics of organised crime that has been accepted by the Council of Europe and NCIS in the UK:

- it is a group activity;
- undertaken for profit;
- involving long-term criminal activity;
- frequently international in character;
- large scale;
- frequently combining both licit and illicit operations;
- involving some form of internal discipline amongst the group, including use of violence and intimidation. (Home Affairs, 1994-5, x)

Such lists of characteristics might be helpful to practitioners (cf. Abadinsky, 1997, 4-8 who also adopts this approach); indeed, variations on this theme are sometimes developed into systematic targeting criteria (see chapter six) but they can be problematic if interpreted rigidly such that activities that are both organised and damaging are excluded from official attention simply because they fail to exhibit one or other characteristic (Levi, 1998, 335).

Criminal organisations

A third way of defining the term has been less in terms of *what* is done or *how* and more on *who,* especially their organisation. For example, Donald Cressey, whose work in the 1960s provided the groundwork and subject for much of the subsequent debate, identified a bureaucratic/corporate model as best describing Italian-American crime in the US. Each 'family' was a hierarchy (boss - advised by counselor, underboss, lieutenants and soldiers) with extensive division of labour and well-understood, if unwritten, rules of conduct. A national commission had overall authority over the 24 US families (Abadinsky, 1997, 9-13). This model has been criticised on a variety of grounds.

For example, Joseph Albini argues that it relied on information from law enforcement (that is clearly gathered for purposes other than academic), the unreliable and contradictory testimony of Joe Valachi and a mis-reading of history (1997, 20-24; also Block, 1991, 2-16). Albini's arguments are energetically challenged by Charles Rogovin and Fred Martens arguing that the validity of Cressey's approach has been borne out by subsequent research (1997, 26-36) but the corporate model remains vulnerable to two different but related criticisms. It overestimates the significance of hierarchies in the organising of crime and, second, did not develop the corporate business analogy in other ways. Specifically, it rested on a separation of 'crime' from 'business' that saw the former as an alien force seeking to subvert legitimate society rather than seeking an understanding of the more complex interaction of the two (Smith, 1980, 368).

The first criticism has always been inherent in the main alternative model of Italian-American crime developed by Francis Ianni that explained it through patrimonial or patron-client networks. These, based on the normal structure of social relations in traditional societies, give a more flexible image of the crime 'families' as a social group attempting to adapt and survive in a hostile environment than does the bureaucratic model (Abadinsky, 1997, 13-27). In time a general consensus emerged among the academic protagonists that neither model could lay claim to any essential superiority and that, given the complexity of crime, no single model could explain all (Kelly, 1997, 39-51; Fiorentini & Peltzman, 1995, 5).

However, although the academic debate as to the best way of characterising organised crime moved on, the imagery of the hierarchical model retains a powerful hold in some areas of law enforcement, at least in part because it is seen to serve the bureaucratic interests of law enforcement agencies in obtaining more extensive legal powers and protecting budgets (e.g. Block, 1991, 22-3; Naylor, 1997, 22-3). Another cause of this is 'mirror-imaging': that those working in intelligence may assume that their targets (nations or groups) think, act and organise in the same ways as they do. Thus, since CIUs are formally situated within hierarchical organisations, they are prone to assume that their targets work similarly.

The main source of empirical work and theorising with respect to organised crime has been the US. It is not only important to avoid the simplistic error of assuming that the phenomenon manifests itself in the same form everywhere but to note that there is no single interpretative scheme that can apply everywhere, especially when the variety of markets within which criminal organisations operate is taken into account (Fiorentini & Peltzman, 1995, 5). For example, in his work on the Netherlands, van Duyne notes the lack of any general 'blueprint' for organising trade, whether criminal or not. He found that the cultural background and upbringing of participants to be important determinants in decisions relating to the use of violence, co-operation with other groups and interaction with the 'upperworld'. So, for example, ethnic minority crime-enterprises are usually family businesses operating with the protective sub-culture, while North-west European crime-entrepreneurs depended less on family and more on networks of friends and connections (Van Duyne, 1997, 204).

Certainly in the UK an approach to the subject based on the study of the interaction of networks and markets is likely to bear more fruit than looking for criminal hierarchies. The vigorous debate that has taken place in the US regarding the nature and significance of the Mafia has not been rehearsed in the UK; no-one has seen the Mafia as a significant player in serious crime in the UK. Although capitalist democracies in the broadest sense, the political economies of Western European countries other than Italy have never provided the conditions within which Mafia-type organisations prospered in the US. But there have been echoes of the debate in the UK. For example ACPO's 1985 Broome Report adopted a hierarchical three-tier model of drugs markets with traffickers at the top, wholesalers in the middle and user/dealers at the bottom. These were to be policed, first, by the National Drugs Intelligence Unit (NDIU) and RCS, second, by specialist force squads and, third, divisional CID (Dorn *et al*, 1992, 209-11). Thus there appeared to be a happy coincidence (caused by 'mirror-imaging'?) between the levels at which both police hierarchies and drugs markets operated. Dorn and his colleagues found that there was no such coincidence:

> Rather, a large number of small organisations operate fairly autonomously of each other in a manner that may be described as 'disorganised crime'. (Dorn *et al*, 1992, 203)

Hobbs argues that the development of organised and professional crime in Britain has been dominated by markets and networks; there may be occasional evidence of elements of more formal organisation but the dominant motif is of localised networks of individual entrepreneurs. To the extent this picture has changed it is because the current domination of drugs requires levels of market organisation that transcend local networks and require more widespread alliances (1994). Hobbs subsequent research also showed that organised criminality is expressed through local trading networks that are flexible, interweaving both legitimate and illegitimate opportunities and vary widely between different areas in relation to local patterns of migration, employment and work and leisure cultures (1998, 407-22).

The House of Commons Home Affairs Committee inquiry into organised crime during 1994-95 concluded in similar vein:

> ...throughout our inquiry it was stressed that attempts to describe or list the various criminal groups would prove over-simplistic: in practice, the picture is complex, with groups of varying degrees of organisation and varying degrees of criminality networking on an *ad hoc* basis as opportunity arises. (1995, HC 18-I, §35)

ACPO itself, concerned in the mid-1990s that NCIS was seeking to establish its credibility by 'ratcheting' up the notion of 'serious crime' beyond those things of routine concern to local forces, carried out its own study. This was not published but the main findings are discussed by Peter Stelfox:

> The picture that emerges is one of individuals or small groups, based within their local communities, acting independently within networks which facilitate their criminal activity. This activity takes place locally, nationally and internationally, and there is no identifiable structural division between these levels of offending. (1998, 401)

Criminal markets

Current discussion at least starts from a position with which almost no-one - whether official or academic - appears to dissent, that is, organised crime is concerned in some way with the operation of markets, either in goods and services that are illegal in themselves such as drugs or prostitution or under-cutting the regulation or taxation that is present in the legal market, for example, tobacco, arms.

The issue then becomes whether criminal organisations are seen as *penetrating* the workings of legal markets or, rather, that legal and illegal markets are seen as *symbiotic*. As ever in social science, what you see depends to a large extent on where you sit. The former image is of honest business persons operating within entirely legal market places being 'made offers they cannot refuse' by people seeking fresh criminal opportunities or outlets for their excess funds. Now, this may be an unhelpful stereotype but if one accepts that it is possible to identify at least *some* difference between predominantly 'legal' or 'illegal' markets, then it is worth examining the variety of reasons why criminal organisations might wish to 'infiltrate' the former. It may be because exercising control within those markets, with their veneer of legitimacy is highly profitable, but there are other reasons. Some criminals may want a legal form of investment as a form of pension, others may be wanting to establish a more reliable way of transferring assets to their biological family or, third, be investing to diversify income sources and thereby reduce risk. None of these necessarily involve the criminal manipulation of the corporations into which the funds flow (Naylor, 1997, 29-30). Petrus van Duyne points to the similarly simplistic notions that govern official policy regarding money-laundering (1998, 359-74).

The image of crime penetrating the legitimate business world derives, in part, from the same distortion in traditional criminology that has produced much greater research into, for example, predatory 'street crime' than into white collar or corporate crime. Official and media concerns with street crime reinforce each other and, *via* research funding, bring about this concentration of academic research. From an alternative standpoint, however, quite a different picture emerges. Let us turn the image on its head and start with 'legitimate business'. In their recent review of the field, Gary Slapper and Steve Tombs seek to explain the prevalence of corporate crime and the weakness of its regulation. They demonstrate that the economic and physical cost of corporate crimes not only exceed those of street crime but that they also exacerbate the very structures of inequality and vulnerability generated by capitalism (1999, 54-84). They suggest that any full explanation of corporate crime would need to take account of:

> the general state of national and the international economy, the nature of markets, industries, and the particular products or services with which particular corporations are involved, dominant ideologies and social values, formal political priorities and the nature of regulation, particular corporate structures, the balances of power within these, the distribution of opportunities within and beyond these, and corporate cultures and socialisation into these. (1999, 160)

This list of factors would serve just as well in generating an explanation of organised crime, for example, as we shall see below, the liberalisation of the international economy has been seen as a major factor in the growth of transnational organised crime and markets are more crime-prone when the goods and services concerned are those for which there is strong demand despite their prohibition or regulation. Evidently, where the dominant ideology is of wealth-creation and entrepreneurship, then the non-availability of purely 'legal' means to wealth-creation fails to dissuade all from seeking illegal ones and, though political rhetoric may erect 'crime-fighting' to a governmental priority, in terms of resources it will have to line up with other policy priorities and selective enforcement will be the rule.

It is not necessary to accept either of these 'ideal-types' - the perfectly legitimate corporation penetrated by the predatory criminal gang, nor the purely criminal corporation, to suggest that a model of 'enterprise' crime and a critical examination of the interaction of *all* markets (whether formally legal or illegal) is more likely to produce insight into this problem than any other approach. For example, Dwight Smith provides an interesting comparison of how 'licit' and 'illicit' corporations deal with their customers, suppliers, competitors and regulatory agencies (1971, 15-29). In later work Smith develops the idea of 'a spectrum based theory of enterprise' in which 'market dynamics operating past the point of legitimacy' is the central context for the illicit entrepreneur regardless of his organisational style or ethnic roots. Drawing on organisational theories Smith suggests this has four main aspects: entrepreneurs, customers, stratification and power. The first two are both situated on a spectrum between the supply and demand for legal products and services in a legal way to the supply and demand of illegal products; this is directly comparable with the horizontal spectrum in Figure 1.1, p.15. The interaction of these two gives rise to the third factor: stratified marketplaces. For example, entrepreneurs may prefer to borrow money from a bank but, if unable to do so, may go to a loanshark; a company with waste to dispose of may, especially during market downturns, take advantage of cheaper rates for illegal disposal. Fourth, the power position of any enterprise depends on its complex interdependence with its market environment and the formal legality or otherwise of the market is only one factor contributing to this position. Building on this Smith suggests that three main images of firms appear: the paragon, the pirate and the pariah. The first two, represent respectively the saintly (always legal) and sinful (always illegal) poles of the spectrum while the third represents those firms 'at the margin of legitimacy' that, in the US, are normally associated with the vice industry (Smith, 1980, 375-83).

Similarly, Vincenzo Ruggiero (1998) discusses how the official economy benefits from the goods or services provided by conventional organised crime, and vice versa and describes the resulting overlap between the 'licit and illicit' as 'dirty economies'. A good example of this is the tobacco industry where there is strong evidence of the collusion of the major manufacturers in the large-scale smuggling of cigarettes (*The Guardian*, January 31, 2000, 1-3). Thus organisations might operate at different points on his spectrum at different times, for example, what Anna Karp has described as state-sponsored 'gray markets' in arms exist as evidence of policies in flux; that is, they may represent diplomatic innovation and testing of the waters (Karp, 1994, 175-89). Alternatively, different subgroups of the same organisation may be situated at different places on the spectrum; as in Matrix-Churchill, this may indicate the circumvention by one part of the state of the restrictions imposed by another.

Criminal governments

Given the emphasis here upon the behaviour of organisations within illegal markets, it is not surprising that the most common analogy applied to the criminal organisation is that of the 'firm'. To the extent that their central focus is on the production and/or distribution of goods and services then the logic of the analogy is clear. However, from a different perspective what is distinctive about 'Organised Crime' and the reason why it is a threat requiring the particular close attention of law enforcement is not that it is a larger or smaller player within markets but, rather, because its performance of certain other functions constitutes a threat to the state's monopoly of legitimate violence - in other words, it is more 'government' than 'firm'.

At a fundamental level it can be seen that states actually develop out of organised crime: Charles Tilly examines the development of European states, specifically the central place of their organisation and deployment of violence. Each of the core state functions - war-making, state-making, protection and extraction - are seen to have developed *via* characteristic forms of organisation and violence. State-making, for example, produces specialised instruments for the surveillance and control of the internal population with a view to eliminating internal rivals, extracting resources and protecting its chief supporters. As Tilly points out, when these activities are less successful and smaller in scale, we call them 'organised crime' (1985; see also Dandeker, 1990, 119-33).

This longer historical view helps us to retain a sense of proportion but we should not be surprised if state officials express concern that a non-state criminal organisation attracts significant loyalty through its distribution, extraction and order-maintenance policies because this represents a challenge to the very process of state-making, to the process by which all states seek to have *their* writ running throughout society. Part of the 'state-making' process is that state and criminal organisations 'compete' regarding both their distributive - provision of goods, services, protection, employment and so on - and their extractive functions. To the extent that the political establishment and criminal organisations derive their support from different sectors of the population, this 'separation of power' can benefit citizen/consumers. What they have most to fear is collusion between the political establishment and crime bosses (Fiorentini & Peltzman, 1995, 15-16); precisely the situation pertaining in contemporary Russia, according to Yuriy Voronin's account of the 'emerging criminal state' in which organised crime and state officials work together. Organised crime, he goes on, 'represents the only fully functioning social institution,' providing many of the services that citizens would normally expect from the state: protection of commerce, providing employment, settling disputes and private security replacing state law enforcement (1997, 53-63). But even in states that have not reached the level of inter-penetration with criminal organisations that the Russian has, this lens of competing regulatory frameworks will be useful for comparative analysis of issues such as corruption.

For Thomas Schelling what was distinctive about 'Organised Crime' compared with 'crime that is organised' was that it sought exclusive authority in an area:

> 'Government' would come closer to what I have in mind. The characteristic is exclusivity, or, to use a more focused term, monopoly. From all accounts, organized crime does not merely extend itself broadly, but brooks no competition. It seeks not merely influence, but exclusive influence. In the overworld its counterpart would be not just organized business, but monopoly. (1971, 73)

Although there are some markets which by their nature are more amenable to a higher degree of central co-ordination than others (Fiorentini & Peltzman, 1995, 5) the problem with a monopoly model of criminal organisation is that the 'span of control' in illegal organisations is relatively narrow - the need for secrecy, the lack of formal, enforceable contracts of employment etc. all contribute to this. To be sure, the organisation develops its own ways of dealing with this problem, including violence against errant members and information control such as loyalty codes (cf. Cohen 1977, 105) and to this extent Mafia-style crime families can be seen less as business organisations than as 'underground governments'. Structured much more as a 'patron-client network' than a hierarchy (Abadinsky, 1997, 19), members and associates have wide 'license' to engage in marketplace activities. The governing role of the Boss includes regulating disputes, conducting 'foreign relations' with outside authorities, and providing social security and protection functions (Naylor, 1997, 15-6; cf. also Fiorentini & Peltzman, 1995, 14-6; Vertinsky, 1999, 179). In return, of course, the Boss, would 'extract' taxation in the form of tribute.

Now it can be seen how this might provide a 'parallel' government to the state in the same way that the 'illegal' economy parallels the 'legal'. As such, arguing 'ideal-typically', *any* successful claim by an organisation to loyalty prior to the state that is backed up by the use or threat of force constitutes a challenge to state legitimacy. But does this necessarily constitute a serious threat to the state? Cressey argued that Cosa Nostra was 'an illegal invisible government' that sought to *nullify* the government through bribery. But while criminal entrepreneurs certainly want to manipulate and use government they much prefer to have a government than none at all because governments provide a degree of market stability and can be used to repress unwelcome competition (Smith, 1971, 24-5).

Transnational organised crime

The concept of organised crime has been around, at least in the US, since the 1920s, but that of *transnational* organised crime has developed within the last ten years. The phenomenon itself is not that new, for example, cross border smuggling has existed for as long as nation-states (e.g. Lupsha, 1996, 23) but a number of factors have coincided to bring it to prominence.

The United Nations Conference in Naples in 1994 has been described as a 'watershed' and a year later the G8 Conference in Canada set up an Experts Group of legal and law enforcement officials (Clay, 1998, 96). But the issue permeates beyond law enforcement ministries: when the G8 finance ministers held their pre-summit meeting in Birmingham in May 1998 half of their communiqué concerned initiatives aimed at financial crime. It is not surprising that officials regard this as a massive problem since they are given some extremely large figures that, in the very nature of things, are no more than guesswork - inspired or otherwise. For example, the 1994 UN Conference was told that global trade in illegal drugs was worth $500billion (*Independent*, May 14, 1998, 24) and in the same year James Woolsey, then Director of the Central Intelligence Agency (CIA), said profits from this trade were $200-300billion (Raine & Ciluffo, 1994, 137). Money-laundering experts estimated, we are told, that the global 'criminal product' from organised crime was $1,000billion in 1996 (Clay, 1998, 95). How do they know? Such figures must be taken with a large sack of salt and yet it is clear that some transnational criminal activities can cause significant harms to both public and private interests. For example, the trafficking of people into slavery, and of arms, as well as the violence associated with drugs-trafficking cause much social harm whereas others do not, for example, trafficking in stolen vehicles and plant.

There has certainly been an increase in such activities during recent decades but this is not surprising: as we have seen, almost the only generally-accepted definition of organised crime is that it involves the exploitation and manipulation of markets. The increased rate at which the international market-place has globalised with its concomitant growth in international movements of people, capital and information has therefore intensified the opportunities for profit in all directions, whether legal or illegal.   Therefore it is not surprising that the 'threat' is perceived as 'foreign', for example, the main drug-producing countries are poor and located in Asia and South/Central America while the main consuming countries are rich and industrialised. But 'supply-side' accounts of the drug-trade are one-dimensional and ignore the social and economic conditions within Europe and North America that permit and encourage the consumption of drugs (e.g. Hebenton & Thomas, 1995, 193-4).

Another part of the explanation for this is that in official circles there has always been a strong tendency to view crime as the responsibility of 'foreigners' and thus, at least in part, as a transnational phenomenon. For example, Michael Woodiwiss traces the development of the Federal Bureau of Narcotics in the US and argues that

> The conception of organised crime as an alien and united entity was as vital for the law enforcement community as the conception of communism as an alien and united entity was for the intelligence and foreign policy-making community. (1993, 13)

At times these policies became entangled: for example, the co-option of organised criminals to prevent sabotage in the US docks during the second world war and communist influence in Marseille after the war. 'Crime' and 'security' have been similarly intertwined throughout the periodic 'wars on drugs' in those countries of Asia and Central and South America in which the US declared a national security interest in the nature of the local regimes (e.g. Block, 1991, 212-13).

The growing economic and social insecurities associated with globalisation have been met, perhaps predictably, by governments scapegoating 'foreigners' and the ethnic 'other'. Most contemporary accounts of transnational organised crime are consequently dominated by lists of ethnically-defined groups with little serious effort being made to compare their role with that of indigenous groups. For example, a recent survey of the impact of transnational organised crime on Canada acknowledges at the outset the hardly surprising point that the foremost advantage of Canada for transnational criminal organisations is its provision of unparalleled access to and from the enormous US market, itself the original and largest centre of organised crime. Yet the significance of this and the development of NAFTA (see discussion in chapter two) receive scant analysis compared with what are identified as the *main* consequences of global modernisation - the increases in criminal activity originating from the Third World and Eastern Europe (Rioux & Hay, 1995).

In the UK the same tendency to target ethnic groups is manifest, however large or small a role is ascribed to indigenous groups. The then Director of the UK Division of NCIS noted briefly that 'domestic groups are predominant in organised crime' before outlining over the next six paragraphs the European, Asian, African and American origins of those commonly suspect (Clay, 1998, 95-6). Others more or less ignore indigenous groups, for example, an academic devoted about half his submission to the Commons Home Affairs inquiry to an overview of 'organised crime and subversive groups' (Rider, 1995, para.1) and (on a content analysis of paragraphs, in descending order) considered the following: Chinese (17.7%), Italian (14.5%), East European (12.9%), Caribbean, Japanese, Terrorists (6.5% each), 'Travellers' (3.2% including Outlaw Motor Cycle Gangs, New Age travellers and gypsies), Right-wing extremists, pseudo religious sects (1.6% each). 'Other ethnic' including Turks, Pakistanis, Nigerians and Jews accounted for 17.7%. By comparison indigenous groups took up 3.2% and he noted, without apparent irony, that the issue of their existence had been overshadowed by the discussion of more exotic groups such as the Triads and Mafia (para.69)! He saw few indigenous groups as being involved with enterprise crime and, where they were, noted that the class system had kept them from encroaching on the business world. However, this might not last: recent investigations had suggested that 'the ingenuity and resourcefulness of the British criminal classes' had been 'grossly underestimated'! (1995, paras.69-70)

The Commons Home Affairs Committee reported being told that 40% of those the German federal criminal investigation agency considered to be involved in organised crime were German passport holders and Dutch ministers reported about half the people involved were Dutch (1995, paras.21, 23). On the face of it, this suggests that the perception of foreigners being disproportionately responsible for crime is borne out. But there is a danger of a circular argument here - if organised crime is defined in terms of specific offences, especially drugs, and the source of those drugs is abroad then it is highly likely that some of those with connections to the source countries will find the market in supply very lucrative. Studying the specific market conditions within which illegal activities prosper is far more likely to produce a realistic appreciation of the issue than ethnic stereotyping, with all its attendant dangers.

If the supposed alien origins of the transnational organised crime threat are one echo from the past, then another is its representation as a centrally-organised conspiracy that represents an even greater threat to the Western industrialised democracies than did the Cold War. Not surprisingly, this line comes from the same sources for whom, first, communism, then in the 1980s terrorism represented similar challenges. The Center for Strategic and International Studies in Washington DC published a transcript of its considerations on global organised crime in 1994 under the sub-heading 'The new empire of evil' in which its project director claimed

> Transnational crime chiefs are busy carving up our little planet into privileged sanctuaries for everything from counterfeiting and credit card forgeries to money laundering and the smuggling of radio-active materials as well as illegal immigrants (Raine & Ciluffo, 1994, xi).

The idea of a new *Crime-intern* to replace the *Com-intern* has been given its most publicised expression by Claire Sterling in *Thieves World* and is effectively criticised by Tom Naylor (1995, 37-56). It does seem extraordinary that, at a time when the world is being analysed through the increase in fragmentation, complexity and diversity that those very 'underworld' activities in which centralised co-ordination has always been hardest (because of the narrow span of control over the 'illegal') are said to achieve levels of co-ordination unattainable in 'upperworld'. A more convincing alternative is that it is the very flexibility of criminal organisations working through networks that accounts for their strength and ability to take advantage of shifting market opportunities.

It is also now widely argued that transnational organised crime constitutes a threat to national security itself, for example, in seeking to explain why certain criminal activities have come to be seen as global problems, Gregory suggests that part of the answer is that it is the 'polis' itself that is the potential victim (1998, 136). For example, the Canadian Government directed the Canadian Security Intelligence Service (CSIS) to collect information regarding this threat under its statutory duty to investigate foreign-influenced activities detrimental to Canadian interests. Accordingly CSIS maintains that

> Transnational crime...poses a serious threat to the economic security of the nation in that its basic activities could undermine the workings of the free market economy. (www.csis-scrs.gc.ca 12//11/98)

Similarly James Woolsey when CIA Director said:

> ...the threat from organized crime transcends traditional law enforcement concerns. They affect critical national security interests. While organized crime is not a new phenomenon today, some governments find their authority besieged at home and their foreign policy interests imperilled abroad. (in Raine & Ciluffo, 1994, 137)

Thus since 1991 the readiness of security intelligence agencies such as the CIA, CSIS and the UK Security Service to advertise the utility of their intelligence skills to law enforcement has given further momentum to the growing representation of organised crime as a security threat. In the US the Cold War re-examination of the size and shape of the intelligence community has included consideration of how intelligence agencies' *modus operandi* can be squared with the procedural niceties of the criminal justice process (Joint Task Force, 1994). In the UK Stella Rimington's public lecture in October 1995 attempted to rationalise Security Service interest in the phenomenon of organised crime by describing it (wrongly) as 'comparatively new' and offered 'the same strategic approach, the same investigative techniques' to counter it as had been developed to deal 'with the more familiar threats' (1995,13). This previewed John Major's announcement to the Tory Party Conference the following week that legislation would be prepared to extend the Security Service's mandate accordingly. A related development has been the increasing involvement of military forces in law enforcement, for example the use of navy ships for interdiction at sea (Sheptycki, 1996, 68).

To hear this argument from security intelligence agencies seeking, in part, to find themselves a new post-Cold War role, is not surprising, but there have also been a number of independent statements to the same effect, for example Phil Williams argued that transnational criminal organisations pose security threats at the three levels of individual, state and the international system of states and he links their rise to the crisis of governance and decline in civil society (1994, 106-110). The problem with much of this literature is that the general nature of the security threat is asserted without adequate analysis of just what aspects of transnational criminal activity do constitute 'security threats' and to whom. After all, the operations of perfectly 'legitimate' markets can also cause massive insecurities, for example, food and raw materials markets in producer countries subject to the purchasing power of industrial corporations.

There is a clear danger that the examination of transnational organised crime, whether by practitioners or academics, could degenerate into the same search for foreign-inspired conspiracies that bedevilled Western internal security policy during the Cold War. This danger is enhanced by the movement into law enforcement of the former security intelligence agencies. But the danger is not just one of bureaucratic manoeuvre - it is also one of language. International relations scholars and political scientists were significantly involved in the articulation of national security policies in the US - and, it might be added, the UK. University international relations departments in the US received substantial funding from the Defense and State Departments throughout the 1960s - now these departments are re-orienting themselves towards issues of transnational crime and associated funding is accompanying that shift. The issue is not just one of academics acting as policy advisors, though that may be problematic, it is also 'about the creation of a discourse which constitutes the symbolic reality of political argumentation' (Lawrence, 1996, 44-59). Didier Bigo identified the same process in Europe where academic and official discussion of the 'new realities' - drugs, cross-border crime and so on - take place in terms of categories that emanate in police and security circles and thus tend to reinforce them (Bigo, 1994, 162). Given that criminal justice research has not been immune from a similar tendency (Sanders, 1997, 187), Dick Hobbs was perhaps unfair in blaming political scientists for the moral panic that has ensued regarding transnational organised crime (1998, 407).

Although recent discussion in the UK has been dominated by European issues, it remains important to recognise the continuing influence of American perspectives. What has been called the 'internationalization' of US law enforcement developed alongside anticommunism during the twentieth century but, following the demise of the Soviet Union, the former has become much more significant amidst the 'new world disorder'. Foreign governments have responded to US pressures, inducements and examples to enact new laws regarding drug trafficking, money laundering, and organised crime and have amended procedures to better accommodate US requests for assistance. In contrast there has been relatively little US accommodation of foreign governments, therefore 'Americanization' rather than 'harmonization' fairly describes global evolution of law enforcement since the 1960s, especially regarding drugs (Nadelmann, 1993). Certainly the US law enforcement presence abroad is significant: a recent US State Department survey found that eight US law enforcement agencies employ 1,649 people permanently assigned overseas (CASIS, 1995, 23). With the exception of the DEA these officers tend just to work as liaison with the agencies of the countries concerned but in February 2000 the FBI announced the opening of an office in Budapest that would have five agents working alongside ten Hungarian police. These agents would be armed and empowered to use the full range of investigative techniques. The Hungarian authorities had overcome the normal reluctance of states to compromise their sovereign monopoly in this way, it was reported, because of the concern with Russian gangs operating out of Budapest (*New York Times*, February 21, 2000).

Conclusion

From this review of the debates around the issue of organised crime, it is concluded that a political economy approach is the most useful way of examining the phenomenon. Recurring themes in the literature are that the offences typically associated with organised crime involve the production and/or distribution of prohibited goods and services - that is, the concern is with the operation of 'markets' and the specific concern of practitioners is the possibility or not of eliminating, controlling or regulating those markets. The specifically 'political' side of the approach is evident in a number of ways: for example, because states remain the primary force seeking some degree of market-control (despite their profession in other contexts that 'You can't buck the market'!) and criminal organisations seek to resist that control, then there is a power struggle. Furthermore, to the extent that criminal organisations are defined by their readiness to use violence to enforce their decisions, then they constitute a challenge to states as monopolists of the legitimate use of violence in their territory. In some cases it can be seen that criminal organisations by a combination of provision of goods, services, protection, enforcement and 'welfare' come to be seen as 'states within the state'.

Given the argument in chapter one that police are best viewed as 'regulators', how can we reconcile this with these alternative views of organised crime as 'markets' or 'governments'? At different times and places the precise nature of the relationship will vary but, at a general level, policing can be viewed as part of the process of 'state-making'. Whether 'organised crime' is seen primarily as the operation and regulation of illegal markets or as performing extractive and enforcement functions, it represents (maybe unwittingly) a challenge to the official state's assumed monopoly of law-making and these other functions. How law enforcement agencies seek to evaluate the extent of the challenge posed by illegal activities remains to be seen; we proceed in the next two chapters to examine some of the main local, regional and national law enforcement intelligence structures in the UK and North America respectively.

4 Development of Intelligence-Led Policing in the UK

Introduction

It was argued in chapter one that information and, in a loose sense, intelligence have always been important yet under-researched aspects of policing, yet they have received more explicit emphasis from practitioners in the UK since 1993. Now, the notion of 'intelligence-led' policing (ILP) is a central feature of current police efforts to be seen to provide 'value for money' and greater effectiveness in the 'control of crime'. In this chapter we discuss the factors leading to the adoption of this policy in the UK and the developing organisational structures for police intelligence.

Until the 1990s Western nations in general defined their feelings of 'insecurity' primarily within the terms of the Cold War; the end of that 'War' has provided more space for other concerns. In some quarters the variety of 'non-state' threats to national and international security are described as 'gray area' phenomena and worry is expressed that the current organisation of states is ill-equipped to counter them (Holden-Rhodes & Lupsha, 1995). At one level the official perception is of problems involving police co-operation across borders whether they are 'horizontal' force borders within, say, the UK (Porter, 1996) or 'vertical' as in federal systems such as US, Germany or Canada.

As we saw in chapter two, the construction of unified markets *via* agreements such as NAFTA and the EU represent an effort by states to manage better the impact of global economic uncertainty but they also set up profound contradictions as de-regulating and extending some markets proceeds alongside greater prohibitions in others. In Europe, for example, establishing the single market in order to maximise the unrestricted movement of people and goods has been accompanied by additional means of monitoring those internal movements (Bunyan, 1993, 15-36).

The concept that has come to summarise these concerns is that of 'transnational organised crime' that was discussed in chapter three. Because of its collective and continuous nature, intelligence methods are seen as especially applicable: for example, the UK Home Affairs Committee noted:

A major recurring theme in the evidence we received was the critical role in the fight against organised and serious crime played by intelligence (1994-5, para.53).

Yet, especially in the UK, the new emphasis on intelligence relates not just to 'organised and serious' crime; it has been presented as the way forward to more effective crime-control at all levels.

The search for new police strategies

At an institutional level the shift from government to governance has manifested itself in attempts (not always successful) to reduce public spending, privatise state assets, 'hive-off' former state functions to private agencies and apply private sector market-driven financial and management techniques to those parts of the state sector that can not actually be sold off. This 'new public management' (NPM) is often based on little more than an ideological belief in the superiority of the market over the state (Walsh, 1995 provides a full discussion). Until the mid-1980s the police were relatively unscathed by the cuts then applied to the rest of the public sector, not least because of their front-line role in the government's battle with the National Union of Mineworkers during 1984-5 but, thereafter, the government became increasingly frustrated at the police's failure to restrain an apparently inexorable rise in recorded crime. Increasing questions about police effectiveness were asked and in 1991 the Sheehy Inquiry, consisting entirely of business persons, was established. The police successfully fought off the more radical proposals in the Sheehy Report (1993) but the subsequent Police and Magistrates Courts Act 1994 demonstrated that the police's former blanket immunity from structural change and the 'managerialist' ethos existed no longer (Leishman *et al*, 1996, 12-17; Savage & Charman, 1996, 45-7; Sullivan, 1998).

A shift in police strategy was given further impetus by the discrediting of the model of crime clearance that had apparently received statutory legitimation in the Police and Criminal Evidence Act 1984 (PACE). By extending police powers, arrest moved from being deployed with respect to specific and relatively serious offences to being used whenever the police found it convenient (Robertson, 1989, 27) and, indeed, police use of arrest increased significantly from 1986 onwards (Hillyard &

Gordon, 1999). PACE also covered in detail conditions and time limits for interrogations which, given contemporaneous (and, soon tape-) recording of interviews, were intended to improve crime-clearance *via* admissible confessions. Apart from the fact that this new model had little impact on clear-up rates, what really finished it off was the series of miscarriage scandals that started with the release of the 'Guildford Four' in 1989. By the time the Runciman Royal Commission (set up after the release of the 'Birmingham Six' in 1991) reported in 1993, the search for some alternative was more or less complete, for example, in 1990 the Metropolitan Police was already claiming the success of a 'new' co-ordinated surveillance and intelligence operation against street robbery in North London (*The Guardian*, June 5, 1990, 3).

The Audit Commission's Report (1993) into police effectiveness identified three main problems with the organisation of policing: the lack of an integrated approach to crime, the failure to make the best use of police resources and the fact that the focus of investigations was on crimes rather than on criminals. Finding the status and resources of police intelligence to be meagre, the Audit Commission recommended that the police increase the proactive element in detective work by developing the targeting of 'known criminals' especially by means of making greater use of informants and other intelligence techniques such as crime pattern analysis and Home Office research projects were established to evaluate and reinforce the new strategy (e.g. see Maguire & John, 1995; Maguire, 2000, 315-36).

Here was the new model of crime-clearance to replace the discredited arrest-confession strategy. In response to perceived *patterns* of crime rather than crime incidents potential suspects would be identified. Information would be sought from informants and other information gathering and surveillance techniques be deployed in the belief that these would provide sufficient evidence of crimes committed that arrest would not need to be followed by a confession in order to obtain a conviction. Of course, confessions would still be sought as insurance and interviews conducted with a view to clearing other offences but 'intelligence' would replace interrogation as the key to getting 'results'.

The attraction of ILP in these circumstances was clear: it responded to the Audit Commission Report in such a way that police could hope to demonstrate a more effective targeting of scarce resources and it provided a new model for crime-clearance that implied less reliance on the now-discredited 'confessions culture' (Condon, 1993). Gary Marx has similarly noted how undercover police work became more important in the US as police saw it as a means of getting around court-imposed and other restrictions on investigative methods such as the 'third degree' (1988, 47). To the extent that ILP was to be directed at dealing with crime, it fitted in well also with police's preferred self-image as 'crime-fighters'. Thus it could meet the needs of government (more effective use of resources, reduce crime rates) and police (restore image as crime-fighters).

ILP could also be seen as contributing towards other contemporaneous developments: the use of local crime pattern analysis (Read & Oldfield, 1995), experiments with 'problem-oriented policing' (Leigh *et al*, 1996, 1998) and local multi-agency community safety initiatives can all be seen as dependent upon a more effective utilisation by police of their information and the development of 'intelligence'. But innovations in policing come and go, there is considerable cynicism among police officers about management innovations and the structures of policework (its low-visibility and extensive discretion) mean that rank-and-file officers are in a very strong position to resist them. The police spend a low proportion of their time actually involved in crime control and the more cerebral demands of ILP stand in some contrast to the primarily hedonistic and action-oriented preferences of police officers (Reiner, 1992, 111-4). Further, police information systems have limited rationality given the emphasis in the police culture of 'craft knowledge', secrecy and an unwillingness to share information (Manning, 1992, 369-72). So we must consider whether ILP is transforming policework or whether it is just the latest fad that is being 'bolted-on' to existing structures. Is the increasing application of ICT actually making policing more 'intelligent' or 'knowledgeable' or is it simply adding to the information overload? (Tremblay & Rochon, 1991).

Complementary to these shifts in public policing has been 'the rebirth of private policing' (Johnston, 1992). Les Johnston has shown how 1829 was not a clear break between 'old' and 'new' police because private police provided the nucleus of the post-1829 public police and continued to be significant long thereafter. At present, he suggests, the balance between the two is being renegotiated through four processes: the expansion of private security, the privatisation of public police; the growth of 'hybrid' policing (cf. Figure 2.1, p.39) and the development of 'active citizenship' (1996, 59-63). The forces driving this renegotiation are, of course, largely those already considered - limited police budgets, beliefs in the superiority of the market and the impenetrability of the private sector to the public police. Whatever the precise causes, however, the point is that policing today involves crucial interactions between these different police sectors and, in terms of the present discussion, the overwhelming majority of those interactions involve not some form of joint *action* between different agencies but exchanges of *information*. Joint actions are relatively transparent, and are therefore more likely to raise controversial questions in the eyes of authorities and public whereas informal information-exchange is relatively opaque.

The organisation of criminal intelligence

Outside of the specific area of political policing (Metropolitan Police's Special Branch was set up in 1883) attempts to develop more systematic intelligence structures began only in 1945 and, as between the different levels of policing, reveal a process of uneven development that is explained by the essentially *ad hoc* and pragmatic nature of UK police organisation.

Local level

Locally, formal intelligence structures came about in the late 1960s with the introduction of unit beat policing. ACPO Guidelines for the development of local intelligence systems were first promulgated in 1975 in the (unpublished) Baumber Report. 'Collators' based in each sub-division represented the police's attempt to compensate for the reduced flow of information from the public as routine foot patrolling was replaced increasingly by vehicle patrols and various specialist squads. A study of a Scottish force in the late 1970s provides the only systematic discussion of this developing system and the way in which, as unit-beat degenerated into 'fire-brigade' policing, the collators acted as conduits of information from the beat to the specialist squads and, apparently, provided little in return to patrol officers (Baldwin & Kinsey, 1982, 74-82).

An indication of how the system was supposed to work is provided by the following:

> The role of the resident beat constable is in helping...to increase and improve the flow of information, thereby raising detection rates...To obtain information, it is useful for the area constable to make

himself known to local officials, shopkeepers, tradesmen, garage proprietors and other reliable persons who regularly visit or reside in each road or street in the area. He should aim at having a contact who is confident in him in every road and street...Information collected should be given to the Collator, who will then pass it to the appropriate department and file it for easy reference....The amount of information passed to the Collator by the area constable will indicate his effectiveness (Merseyside Police, 1983, paras.35, 43, 47, 47).

Meanwhile, the two collators per Division were to record and index 'all items, however insignificant' and ensure they were passed on as appropriate. The collators would maintain a series of indices, for example, regarding vehicles, modus operandi, and 'Beat-Street' - this last containing all information known regarding all addresses on the Beat. The core was the Intelligence Index containing details of 'all local criminals or persons suspected of being active in crime', not surprisingly, but also 'details of all persons who have come to notice' (Merseyside Police, 1983, paras.50-66). Campbell and Connor provided a detailed critique of the developing use and computerisation of such information since much of it related to issues of lifestyle including sexuality and amounted often to little more than local gossip (1986, 190-225). Baldwin and Kinsey were told that well over half the subjects of the equivalent index in their research force had no criminal records (1982, 61). It is clear that patrol officers made little use of the collator's card indices - the Merseyside Police Officers Survey found that less than 1% of uniformed patrol officers time was spent consulting the collator, while CID officers spent only 1-2% of their time (Kinsey, 1985, 60-62) consulting what was, after all, the institutional 'store of memory'.

Current developments seek to place intelligence much more at the heart of local policing. In line with the recommendations of the Audit Commission Report (1991) responsibility for providing local policing has been devolved to the level of the Basic Command Unit (BCU) - more or less equivalent in size to a division. The central feature of the Crime Management Model now being widely implemented is that the BCU manage all their resources - uniform, criminal investigation, traffic, specialist - proactively in accordance with intelligence-led objectives rather than allowing them simply to react (Barton & Evans, 1999, 8-9, 22; Leigh et al, 1998).

> The (Intelligence Unit) is central to any proactive model...It is responsible not only for generating intelligence itself but also for developing strategies and tactics for other teams within the BCU to use against the analysis of problems. In addition the (Intelligence Unit) is responsible for crime prevention (Amey *et al*, 1996, 4).

Thus, intelligence is intended to be used in two main ways, first to target specific 'criminally active' people with a view to developing the evidence necessary for a conviction and, second, to inform crime prevention strategies *via* the analysis of problems.

In one of the research areas in North Wales, the Crime Management Unit was headed by a Detective Chief Inspector (DCI) although day-to-day operations were headed by the Detective Inspector (DI). The 'Crime Desk' into which all reports of crimes come, is headed by a Detective Sergeant (DS) and staffed by four constables. The Intelligence Unit, made up of a Local Intelligence Officer (LIO) and a Field Intelligence Officer (FIO) is situated in a small office adjacent to the Crime Desk. The Crime Prevention and Drugs Prevention Officers were also part of the Unit though located in separate offices. In another research area the terminology was different - the overall structure headed by a DCI was called the Divisional Strategy Unit. Within this the CIU again consisted of a FIO and LIO and the Crime Desk, Crime Prevention and Drugs Prevention officers were in adjacent rooms. Here, however, there were two DS, one responsible for tactical intelligence and the other for strategic intelligence. The former was responsible for developing information on the eight target criminals identified by the area Tasking and Co-ordination Conferences and was specifically assisted by the LIO and the FIO. The latter was responsible for the identification of crime problems, for example, repeat victimisation, repeat calls and was assisted by the 'analyst', a constable who used the call-logging system to identify crime patterns.

Force level

Formal intelligence structures emerged first in the Metropolitan Police, the largest force in the UK. In terms of information gathering the beginnings can be seen in 1945 when the Special Duty Squad was established with four officers working undercover in what became known as the 'Ghost Squad'.

In response to the postwar crime wave, much of it concerned with circumventing the rationing restrictions, they were mandated to make as much use as possible of informants and to pass on the resulting information to the Flying Squad or divisional CID. Despite its claimed successes it was wound up in 1949 (Gosling, 1959, 13-22). A Criminal Intelligence Branch was established in 1960 to co-ordinate information on 'organised crime' and 'prominent criminals' based in London but operating in the provinces (Jackson, 1967, 132-3; Dorn, 1992, 153). By the 1970s this branch had a staff of fifty (Bunyan, 1976, 80). In the mid-1990s the Met's Directorate of Intelligence (SO11) included a Drugs and Violence Intelligence Unit that co-ordinated intelligence-led operations against organised crime in London and thus was involved in much liaison with NCIS and the South East Regional Crime Squad. Each of the Met's five areas had a Force Intelligence Bureau (FIB) and each of the 63 stations an 'intelligence cell' (Smith, 1996, 240-1). At Scotland Yard, the Met's headquarters, the most recent change, and one that again indicates the increased merging of security and law enforcement intelligence, is the formation of an Intelligence Development Group from the amalgamation of Special Branch, the central CID and the covert operations unit (*Intelligence*, N.103, September, 1999, 12).

Further central intelligence units were set up in 1963 in Birmingham, Cardiff, Durham, Glasgow, Liverpool and Manchester and some form of force units became more generally-adopted from the mid-1970s (NCIS, 1998b). In the wake of ILP, most forces have now adopted the model of a Force Intelligence Bureau. For example, in Merseyside the FIB was formed in 1994 and when a Major Crime Unit was established in 1996 to merge the previous squads for serious crimes and drugs, the FIB re-configured its organisation to target the same three organised crime syndicates (Barton & Evans, 23-4). In North Wales, 'level one nominals' (those responsible for volume crime within a particular area) will be the responsibility of the Area Command; those who are identified as 'level 2 nominals' are targeted for collection plans by the FIB *via* the Force Tasking and Co-ordination Group (TCG). These people will be those operating 'at a higher level' of criminality, across divisional or force boundaries or whose apprehension would require more than divisional resources (North Wales Police, 1996, para.2). In 1997 the FIB was headed by a DI and included eight people, including a civilian Senior Intelligence Clerk, a civilian working on telephone analysis and another on 'itinerant' crime.

Regional level

Regional criminal intelligence offices (RCIOs) were set up in 1978 to work with the regional crime squads RCS). Prior to the changes brought about in the Police Act 1997 (see below), the main review and adjustment to the regional system came in the mid-1980s. Triggered in part by the temporary withdrawal of one force from its RCS and threats to do so from others, the Home Office Review recommended that RCIOs and the Technical Support Units (TSUs - regional stores of specialist surveillance equipment) be incorporated into the RCS funding and organisational structures (Home Office, 1986, para.13). The Review also took into account the creation of separate drugs wings attached to RCS in 1985 but regional intelligence work remained integrated as between drugs and non-drugs matters (Dorn, 1992, 154). The nature of the targets of regional policing meant that policing there always displayed greater elements of intelligence work. The specific work of RCIOs would take place on more future-oriented work, as projects moved nearer to fruition in terms of an operation then they would withdraw and more RCS people would become involved as 'intelligence' gives way to 'evidence' gathering. The North West RCS itself was organised into three main offices and two smaller ones. Each of the former had a staff of 55 including a Development and Research team with four to six police or civilian analysts working on, for example, telephone billing. The latter had one or two 'syndicates' - seventeen-strong surveillance teams but were without specialist analysts (Nicholls, 1996).

The RCS represented a formal organisational response to the issue of cross-border crime but, even before their upward amalgamation into the National Crime Squad, they could not deal with all those issues. For example, interforce borders might also be inter-regional borders, or forces might agree that they wanted to deal with a problem that would not be accepted by the RCS, for example, if it failed to meet its targeting requirements. One way in which such co-operation might be furthered, according to the Home Office, is through Regional Tasking and Co-ordination Groups consisting of representatives from forces, RCS, customs and NCIS (Porter, 1996, 24) but comments from police officers at a conference on 'Cross-Border Crime' held in Manchester in January 1997 suggested that such Groups lacked the necessary resources (Author's notes). More likely to be productive from the point of view of practitioners are informal networks that develop between officers and agencies perceiving a common organisational interest in co-operating. Examples of these would include the 'Burglary Artifice and Itinerant Conference' formed, first, by police in the South and East Anglia and then by police elsewhere to cover the Midlands (Porter, 1996, 25-6).

National level

In 1975 the Baumber Report recommended the establishment of regional and national criminal intelligence offices; the latter proposal was repeated in the Ratcliffe Report (1986). The Central Drugs Intelligence Unit that had been set up in 1972 became the National Drugs Intelligence Unit in 1985 (Dorn, 1992, 153-4) and was then integrated with other non-drugs national intelligence units to form the National Criminal Intelligence Service (NCIS) in 1992. NCIS absorbed the five regional RCIOs with their 224 staff and presaged the incorporation of the RCS into the National Crime Squad (NCS) when both national squads were placed on a statutory footing by the Police Act 1997.

NCIS is a multi-agency law enforcement body: in 1996 its staff of 536 was made up of 28.4% provincial police, 16.8% Metropolitan police, 8.2% Customs and Excise, 35.4% Home Office civilians, 2.8% civilians from the Metropolitan Police and 8.4% from local authorities (NCIS, 1996, 27). The three main divisions into which it is organised now are for the UK, International and Intelligence Development. Most of the staff work in the UK Division: 42% in the five regional offices (north-east, north-west, midlands, south-east and south-west) and 15% in the Strategic and Specialist Intelligence Branch. This is organised in 'crime' areas such as 'specialist' (football, paedophiles, kidnap/extortion) 'drugs and organised' and 'economic'. Nineteen percent work in the International Division, the largest section of which is the UK National Central Bureau of Interpol. The largest branch in Intelligence Development (16% of all staff) is 'special projects' that includes units to assist police in applying for interception warrants, for liaison with the Security Service and to manage the National Informants Database (adapted from NCIS, 1999g). Additionally, a number of other agencies maintain liaison officers at NCIS, including the Benefits Agency, Immigration Service and Security Service (Clay, 1997).

The official aims of NCIS are threefold: to process intelligence (including responsibility for all stages of the intelligence process and liaison with other agencies in the UK and abroad), to give direction (by setting standards for users, establishing nationally agreed systems and seeking the avoidance of duplication by law enforcement agencies), and, third, providing services and strategic analysis (including being the lead agency for Europol and Interpol and providing strategic analysis for government and other agencies regarding desirable changes in policy and resource allocations (NCIS, 1996, 29).

It is clear that NCIS' ability to fulfil these aims was limited initially by a number of factors, including an inability to build up longer term expertise through only taking officers from police and Customs on secondment for 2-3 years and, prior to 1997 only being able to recruit from the Home Office pool of generalist civil servants rather than the specialists it wanted (Clay, 1997). NCIS also faced the collective suspicion of 43 chief constables that it represented another move toward the 'nationalisation' of policing and that its constant 'banging the drum' about organised crime represented (an understandable) need for NCIS to mark out its turf but at the expense of that 'local' crime that remained their primary concern (see also Stelfox, 1998). It is clear that NCIS' original objectives were too ambitious given their limited resources but their effective withdrawal from operating at the regional level in order to concentrate on the 'top tier' of criminality frustrated local forces (HMIC, 1997, 1). One of NCIS' predecessors, the NDIU, had failed to live up to other agencies expectations of its ability to deliver up-to-date intelligence (Wright *et al*, 1993, 65) and early criticisms that NCIS intelligence 'packages' similarly failed the test of 'timeliness' further aggravated its credibility problems. Packages were abandoned and the *modus operandi* became the development of joint collection plans with RCS and Customs. However, NCIS' credibility problem clearly had not been solved by 1997 when, according to reports, there was a disappointing quality of applicants to become the head of NCIS in succession to Albert Pacey (*The Guardian*, August 14, 1997).

NCIS, in its earliest days, faced not just pressure from below, as it were. As the Cold War thawed, the Security Service emerged from the chill to offer its services and skills on a variety of areas of police turf. The first of these did not affect NCIS directly insofar as it was in 1992 that the Metropolitan Police Special Branch lost its historical primacy in operations concerned with republican paramilitaries to the Security Service. The new arrangements hardly had time to bed down before the first PIRA cease-fire was announced in 1994 and, once again, the Security Service seemed to be in danger of losing business. This was averted after a period of skilful lobbying in Whitehall that culminated in the Security Service Act 1996 extending the mandate of MI5 to include also 'serious crime' (For a detailed critique of these developments see Gill, 1996). In fact, this was not the beginning of the involvement of security intelligence agencies in 'crime-fighting': in 1986 the Secret Intelligence Service (SIS) decided that it could make a specific contribution to developing intelligence on the destabilising impact of drugs trafficking in some of the dependent territories in the Caribbean. Since then both SIS and Government Communications Headquarters (GCHQ) - the UK agency responsible for the collection of signals and communications intelligence - had been tasked regarding drugs by the Cabinet-level Joint Intelligence Committee (JIC) (Warner, 1998).

However, the specifically domestic rather than foreign remit of the Security Service, and their historical autonomy, meant that the police were very concerned to retain 'police primacy' in areas of unambiguous 'criminality' even if it had lost it in 'terrorism' - an area of more ambiguity. NCIS's strategic aims were drawn up in consultation with ACPO and the Home Office while the autonomy of MI5 has been marginally reduced in that its strategic operational objectives are now in line with JIC requirements (Warner, 1998). When it comes down to actual tasking, matters seem somewhat unclear: formally, at least, NCIS chairs the committee that tasks all three security intelligence agencies, though all three only accept tasks in the context of their other priorities. It is understood that in 1997 the Security Service had only fifteen to twenty people working on matters of transnational organised crime and that they confined their work to areas where they could make a particular contribution, for example, where specific language or surveillance skills were needed. But a report in 1999 indicated that the commitment has increased rapidly - to £10million spent on 'fighting serious crime' in the previous year (*Sunday Times*, December 5, 1999). Another, less public, motivation for empowering MI5 in this way was apparently the concern with police corruption, especially in the Metropolitan Police, and the belief that the Service's surveillance skills along with the fact that it was not itself part of the police, would be more effective than the police's own anti-corruption units (*Sunday Times*, April 9, 2000).

There must be doubt that NCIS is able to translate its formal position into real primacy. The NCIS budget (£34.9m in 1998-99) has failed to increase at the same rate as the rising tide of ministerial rhetoric concerning the threat of organised crime: a former head of the UK Division has written: 'A (NCIS) with fewer resources than those allocated to, say, traffic policing in a large police force is clearly an anomaly' (Clay, 1998, 98). Apparently, following warnings from the Home Office that the UK was facing a 'crime emergency', Tony Blair convened a meeting at Downing Street at which he was briefed by the heads of the three intelligence services and agreed to their request to divert further resources from counter-espionage and counter-terrorism to 'serious crime' (*Sunday Times*, December 5, 1999). The press report of this 'summit' failed to mention whether or not NCIS was present; if it was not, then 'primacy' hardly describes its location in the Whitehall pecking order.

Customs and Excise

We have seen already references to the involvement of Customs in the developing intelligence structures. Its role is significant because its enforcement role regarding trade prohibitions and restricted goods places it at a key point in the policing of drugs markets. It also remains an independent prosecuting authority. Customs has a budget of £842m and 23,000 officers and, as part of the New Public Management, carried out a Fundamental Expenditure Review in the early 1990s. Prior to this Customs was organised in 21 Collections (regions) each of which had an investigation unit. Specialist intelligence officers were not part of any intelligence structure and worked in support of the investigation units. In the customs and prohibited goods side of the agency's work there were, even then, studies relating to the vulnerability of specific ports, but in general, and especially on the revenue side (VAT and excise), most investigation was organised around regular cycles of visits (Wesley, 1997).

As a result of the expenditure review, the number of collections was reduced to fourteen and the investigation service was centralised into one national unit running in five regions, not, as before, based in the Collections. Although intelligence was reduced from a 'core' to a 'non-core' activity, Customs convinced ministers that more resources rather than less should be invested therein. Now the National Intelligence Division (NID) is responsible for intelligence policy, planning, performance measurement and co-ordination. Sixty of its current 160 staff work on the development of ICT. It is also now seeking to provide strategic intelligence and aims to have thirty strategic analysts, the current vehicle for this work being the production of risk and threat assessments (Wesley, 1997). Actual operational intelligence remains the responsibility of the collections. Each collection has approximately sixty staff working on intelligence matters, divided equally between fiscal and smuggling areas. At the core of these staff is the Intelligence Co-ordination Unit that, in the North West Collection consists of fourteen people: two analysts, six in a collation unit acting as the central point for the receipt of information and five Intelligence Officers working operationally.

As with police, Customs too now claim to be targeting their efforts more effectively, for example, by relying less on cycles of visits to traders and preparing risk-assessments to identify the possibly non-compliant and then concentrating resources on them that also causes less disruption to those perceived to be compliant traders. Other changes have contributed to a similar shift at ports: the coming of the single market in 1993 led to the end of officers looking for 'cold pulls' at airports and replacing it with 'intelligence driven' risk assessments and profiles. Thus Customs represents an intelligence structure that parallels the police, it is clearly much smaller and does not have such a broad law enforcement remit as police but the proportion of its staff devoted to intelligence matters (about 4%) is probably higher than for the police.

To the extent that each agency has specific legal responsibilities (trade prohibitions or general law enforcement) for different parts or stages of the same illegal markets (especially drugs) co-operation is clearly central to effective law enforcement. However, precisely these overlaps represent in-built structural contradictions that will militate against co-operation on crucial occasions. For example, ILP in general has led to a growth in the use and targeting of informants who are, by their nature, difficult to control and pose a variety of threats to the agency concerned. This is yet more complex when more than one agency is involved; say, if a person working as an informant for one agency may be under surveillance and/or arrest for their activities by the other. Such conflicts of interest can have messy consequences even if they are not all as spectacular as that involving Customs, RCS Nos. 2 and 4 and Cleveland Police in the case of Brian Charrington, several major drugs importations and the eventual involvement of the Attorney General's office in lobbying around disposition of the case (see Rose, 1996, 174-88 for account). While procedures may be in place to see that such conflicts of interest do not occur, the realities of policing illegal markets mean they are inevitable.

Northern Ireland

It would be misleading to write even a brief history of police intelligence in the UK without making reference to the distinctive contribution of developments in Northern Ireland, especially since 1969. It remains to be seen just how influential these will prove to be in the longer term over what happens in the rest of the UK but the signs are that, should the peace process be maintained, at least certain aspects of the intelligence structures erected there are less likely to be dismantled than re-directed towards new targets.

First, what is the structure that has developed in Northern Ireland? Despite official secrecy, it is possible to construct a broad account of this thanks to the work of journalists, occasional court cases and document finds[1] The intelligence structure in Northern Ireland is a three-pronged affair in which police, security intelligence and military have provided proportionately the most intensive surveillance of the population (1.5 million) anywhere in Europe.

The Royal Ulster Constabulary (RUC) Special Branch ('E' Department) has been the main police department concerned though at various points efforts have been made to break down their self-imposed isolation from the rest of the force. For example, when Kenneth Newman took over as chief constable in 1976 he brought from London the idea of targeting and sought to establish a new criminal intelligence system to support the work of a new Headquarters Crime Squad drawn from both CID and Special Branch. When Jack Hermon took over in 1980 he believed this squad had in turn become a 'force within the force' and imposed a greater circulation of officers (Ryder, 1989, 150, 232). The main sections into which Special Branch is divided are: E1 administration, E2 legal, E3 intelligence (E3A Republican, E3B Loyalist, E3C Left), E4 operations (E4A watchers, E4B technical, E4C/D specialist photographic). Working closely with E4 are the Headquarters Special Mobile Units used for both public order and covert operations.

The second prong is provided by the Security Service, almost half of whose resources have been devoted to Northern Ireland and connected operations in Britain since 1993. It has about 70 officers actually serving in Northern Ireland and one of its Deputy Director Generals serves as the Director and Co-ordinator of Intelligence at Stormont advising the Secretary of State for Northern Ireland. In 1994 the then Co-ordinator, John Deverell was one of the 29 police, military and RUC intelligence officers killed in the Chinook crash in Scotland (*The Guardian*, June 4, 1994, 1). MI5 personnel tend to be attached to specific police or army operations as and when their specific expertise is appropriate.

The third prong is the army. Broadly speaking, one set of intelligence specialists are integrated into the mainstream of army operations while another serve in units that appear to enjoy a high degree of autonomy from the first. The first group includes a Weapons Intelligence Unit, a Joint Surveillance Group that is responsible for the management of the product of the multiple computer systems being used (see further below), air force and Army Air Corps providing aerial reconnaissance, and the Field Reconnaissance Unit (FRU) based at army headquarters at Lisburn and acting as a centralised handling group for informers. Special Military Intelligence Units provide for liaison with RUC Special Branch and have officers at police headquarters as well as about thirty officers in each police division. The principal of this liaison is that the RUC provide the local knowledge and the military provide follow-up action. The main players in the second group are the Special Air Service regiment (SAS) and 14 Intelligence Company. The latter started life as 4 Field Survey Troop and provide covert surveillance; they include also 'Dets' who provide more aggressive information gathering and countering operations.

Given the numbers of agencies on the ground and the lethal nature of the conflict in Northern Ireland in the 1970s, it is perhaps not surprising that the period was characterised by 'intelligence wars' as different agencies competed in terms of their perception of the problem, preferred remedies for dealing with it and sources of information. In an attempt to bring about peace and greater co-ordination of intelligence operations at the end of the 1970s Tasking and Co-ordination Groups (TCGs) were set up in each of three regions (Belfast, South - Gough - and North - Derry) with representatives of each of these three prongs. In recognition of 'police primacy' that had been established in 1976 as part of the overall Government policy of seeking to 'de-militarise' the conflict, an RUC officer was in the Chair. Given the different *modus operandi*, bureaucratic interests and degrees of political autonomy of the three contributors, it is not surprising that this model of hierarchical co-ordination worked fitfully. For example, 14 Intelligence Company is not represented on the TCG (Dillon, 1990, 469) and the FRU's system of informants runs parallel to that of the police. Geraghty concluded that there was no unified intelligence command in Northern Ireland and that if 'synthesis' occurred it was primarily through the SAS or the Intelligence Corps (Geraghty, 1998, 134).

The case of Brian Nelson illustrated very well the longevity of turf wars and the dangers that could result. Nelson joined the Ulster Defence Association (UDA) - the largest loyalist paramilitary group - in 1972, having been discharged from the army, and became an intelligence officer in 1983. He left Northern Ireland in 1985 and went to live in Germany but he returned to Belfast in 1987 having been re-recruited by the FRU as an informer and rejoined the UDA as a senior intelligence officer. However, the RUC were not told of his re-recruitment and MI5 were told but argued against it. Nelson's job included compiling details of Republican targets for UDA gunmen and, in turn, he supplied the army with information as to imminent UDA attacks. His relations with the army were exposed by John Stevens' inquiry into allegations of collusion between security forces and the loyalist paramilitary groups and he was arrested in January 1990. Only some limited part of Nelson's activities and their implications were revealed at his trial in 1992 (*The Guardian*, January 23, 1992, 2; Panorama, 1992).

However, what relevance does this have for broader developments in law enforcement intelligence? Arguably, Northern Ireland represents in microcosm what has happened to western intelligence agencies since the 1970s. As we have seen, the decline in their traditional role of counter-espionage (though it has not disappeared completely) has been filled by, first, counter-terrorism and then by the targeting of transnational organised crime as a 'new' security threat. The military, too, have found new roles, for example, the interdiction of drug smugglers by the navy. In Northern Ireland the official perception of the 'security problem' has increasingly incorporated the issue of financial support for paramilitary groups and therefore has moved beyond political violence *per se* to take into account associated drugs trafficking and assorted 'rackets' around clubs and building contracts. The new Part III of the PTA introduced in 1989 reflected this issue of financial support (Walker, 1992, 109-30). Therefore, there is no reason to suppose that peace in Northern Ireland will witness the 'dis-invention' of methods and techniques developed there. Indeed, in a number of respects, the security situation there has already provided a preview of various techniques that have subsequently found their way into policing more generally in the UK (Hillyard, 1987).

The first are *legal*. Northern Ireland has been governed by special powers since partition in 1921 but the outbreak of political violence in 1969 was eventually met by further efforts to create a legal regime initially quite distinct from that in the rest of the UK. Detention without trial was ineffective in a security sense and seriously counter-productive politically and so the British government switched to a strategy of 'criminalising' the conflict. This involved ending internment but also changing the criminal justice process in such a way that it could be relied on to produce more convictions than a system that was vulnerable to sectarian division and the consequent unwillingness of witnesses to testify and juries to convict. Some, though certainly not all of the changes made have found their counterparts in Britain, for example, the practice of anonymous security intelligence officers giving evidence behind screens. Limitations on the traditional right to silence of a person detained by the police were introduced first in Northern Ireland in 1988 on the grounds that they were required by the special circumstances there, for example, paramilitaries trained in counter-interrogation techniques, but were later introduced in Britain also in the Criminal Justice and Public Order Act 1994 (Card & Ward, 1994, 153-80).

Similarly, police in Northern Ireland have always had greater search and arrest powers than those in Britain but they were also further augmented by the emergency legislation of the 1970s. Also, what we have seen is that changes brought in initially on the pretext of being useful in the policing of 'terrorism', for example, the right of police to stop and search individuals without reasonable suspicion, has then been used in the context of other 'serious' crime. We should not be surprised at this slippage: just as special powers were seen as necessary in Northern Ireland to deal with the specific threat of violence from organised groups of paramilitaries deploying their own counter-intelligence techniques (Hogan & Walker, 1989, 46-85), so, as we saw in the last chapter, the threat of transnational organised crime is seen in much the same way.

In terms of the *organisation* of intelligence, as we saw above, the military have been particularly influential. This doubtless reflects in part the paramilitary nature of the conflict in Northern Ireland but another factor was the lack of any distinctive law enforcement model of intelligence. Whereas the paucity of police intelligence was quickly exposed by the internment fiasco, the military initially applied their counter-insurgency thinking from previous postwar colonial campaigns, for example, 'counter-gangs' (Dillon, 1990, ch.2). Even after the formal establishment of police primacy in 1976, the size of the military commitment in Northern Ireland

and the fact that they and the Security Service provided training for the RUC meant that the influence of military thinking remained high. Some of the military intelligence language from Northern Ireland has been widely adopted by mainland police, for example, the TCG as a co-ordinating mechanism and, arguably, military thinking remains generally influential. As long as the rhetoric of 'wars' on crime and drugs remains, this is unlikely to change. Another organisational development has been the 'co-ordinating interagency' model represented by the establishment of the Terrorist Finance Unit within the Northern Ireland Office in 1989 with police, Customs, Inland Revenue personnel and accountants with the mandate to enforce the new financial provisions of the Prevention of Terrorism Act, 1989 (Norman, 1998, 377-82).

A third factor that reinforces this is *personnel*: the military provides a source of ready-trained intelligence personnel, both for analytical work and surveillance. NCIS, for example, appointed a former military officer as its first head of analysis in 1997. The lack of a career structure for civilian analysts or much by way of training is leading to much poaching from the services (see further in chapter nine). It is understood that soldiers trained in undercover work in Northern Ireland but currently not required because of the cease-fire, have already been 'lent' to the Security Service as watchers and not only in support of terrorist-related operations.

Technological factors are also relevant: clearly, Northern Ireland has provided a testing ground for 'state-of-the-art' technical methods, for example vehicle and face recognition systems. More has recently become known about the main computer systems deployed by the military owing to the IRA obtaining copies of intelligence documents including target files and operational manuals that had been the property of the Welsh Guards (*Republican News*, January 28, 1998). The two main databases related to names - CRUCIBLE - and vehicles - VENGEFUL and efforts have clearly been made to develop their analytical potential, for example, so they can provide link analysis. The latter is linked to the Northern Ireland vehicle licensing office and it was intended to be linked to a network of 100 cameras (eighty in overt, twenty in covert locations) capable of license plate recognition. However, the extent to which the military in Northern Ireland have succeeded in linking the large number of disparate programmes that have developed there (37 in 1994) remains unclear (Geraghty, 1998, 158-61). The technologies of surveillance have clearly been spreading rapidly also in mainland Britain, some have been initiated explicitly because of the perceived threat of terrorism, but the ubiquitous city centre CCTV schemes, have had rather more prosaic goals - making cities safe for consumption. However, this does not mean that they cannot be deployed more generally in the construction of local security complexes (see chapter two).

Finally, there are *tactical* factors. One of the historic roles of the police has always been the 'prevention' of crime, but a key element of the difference between police, on the one hand, and military-cum-security intelligence on the other, is that the former are said to be concerned with the arrest and prosecution of offenders and the latter is not. The latter would be far more concerned to disrupt the activities of its opponents, in some cases this might be relatively benign such as rendering harmless a cache of weapons but, if necessary it would be lethal. Soldiers are, after all, trained to shoot people who are deemed to be enemies, rather than negotiate with or arrest them. There have been a succession of cases in Northern Ireland where, mainly, the autonomous squads such as SAS and 14 Intelligence have 'taken out' paramilitaries in ambushes but the RUC also used such tactics in the series of shootings in 1982 that sparked off the Stalker inquiry (see Stalker, 1988, 253-75 for his own conclusions). As we shall see in chapter nine, there is clear evidence that police are now more likely to use disruptive methods against criminals where they believe that it would be easier or cheaper than pursuing an investigation through to arrest and prosecution.

Conclusion

Steadily, though unevenly, during the last fifty years, specific structures for covert information-gathering and the production of criminal intelligence have developed in the UK. The pace of change increased considerably during the 1990s, in particular because the police themselves embraced the possibilities of the new ICT for the processing and retrieval of their large store of information just as some of their more traditional methods, especially obtaining confessions *via* interrogation became discredited. But ironically, perhaps, just as the police geared up to make better use of their information, other agencies arrived on the law enforcement scene with their longer traditions in the business of developing covert security and foreign intelligence. As the new century begins, what we find is a very crowded law enforcement intelligence scene with agencies operating at all the levels and across the sectors presented in Figure 2.1. There is clearly plenty of work for all - 'insecurity' is such a *leitmotiv* of the contemporary world - and the agencies seek to obtain some leverage over the problem by better deployment of intelligence. Before we examine in detail the processes by which they attempt this, we shall discuss equivalent developments in North America.

Note

[1] Unless otherwise referenced, the following account is based primarily on Geraghty, 1998, 130-68; *Statewatch* 2(5), Sept-Oct 1992; Urban, 1992, 93-8. See also Maguire, 1990, 145-65.

5   Law Enforcement Intelligence in North America

Introduction

Compared with the UK there has been less explicit discussion of 'intelligence-led policing' in either Canada or the US. However, intelligence techniques have been applied there in law enforcement for as long, if not longer, and there is certainly more evidence of practitioner reflection and writing on the issue in North America. On both sides of the Atlantic, intelligence techniques were used earliest in the specific context of political policing - the development of special branches in the UK (Bunyan, 1976, 102-29) and the equivalent 'red squads' in major US cities (Donner, 1990). These persisted through the 1960s and 1970s in the context of the political and industrial unrest of the period while the more explicit application of such techniques to 'crime' developed slowly from the 1950s onwards, more or less in line with the perceived seriousness of the threat posed by 'professional' or 'organised' criminality. From the 1970s onwards there were also a number of experiments with the targeting of repeat offenders in the US (Heaton, 2000, 340-2). In the 1990s, developments in both Canada and New York have even more closely mirrored those in the UK, for example, Eli Silverman's study of NYPD:

> What is new since 1994 is the setting in place of unusual engines of organisational change within the NYPD. This change has been sustained not just by strong leadership but through the dynamics of intelligence-led policing. (1999, 179)

Canada

Police in Canada have been making use of intelligence techniques since the late 1950s (Stewart, 1996, 84-5) and of specific intelligence units since the early 1960s (Prunkun, 1990, 20) but little is known about their operations (Brodeur, 1992). Given the RCMP's size (in 1993 its establishment was

22,632), status as national icon (Walden, 1980), and national 'reach', it might have been expected that it would provide the lead in co-ordinating police intelligence efforts but this was not the case because the apparent national reach of the RCMP disguises the fact that it is directly responsible for the 'low' policing of only 10% of the population, and that is entirely rural. It was only in May 1991 that its own headquarters Criminal Intelligence Directorate became operational as a result of the perception that the failure to develop a strategic intelligence capability had seriously hindered the Force's ability to measure, let alone prevent crime having 'an organized, serious or national security dimension in Canada, or internationally as it affects Canada' (RCMP, 1991a, 1; see also Smith, 1997).

Thirty years earlier the first moves towards co-ordination had begun in Ontario. Following statements made by a retired RCMP Commissioner concerning organized crime and the discovery of corruption in the Ontario Provincial Police (OPP), the Ontario Police Commission was created in 1962 to encourage inter-agency co-operation between the province's police forces and the Roach Royal Commission (1961-3) further investigated allegations relating to illegal gambling and organised crime (Beare, 1996, 141). The Police Commission reported further on this in 1964 and concluded that a continuous source of criminal intelligence on organized crime was required; from this the Criminal Intelligence Service Ontario (CISO) was set up in 1966 to act as a central clearing house for criminal intelligence.

In the same year the federal government convened a conference of federal and provincial attorneys general who made two proposals: the establishment of CIUs by all forces with organized crime problems and the creation of a central clearing house such as CISO in each province. Currently nine provinces have done so, and in 1970, after several years gestation, Criminal Intelligence Service Canada (CISC) was established to provide a central facility for the sharing of intelligence between Canadian police agencies. CISC's Central Bureau is based in RCMP headquarters in Ottawa and is made up entirely of police, mainly from the RCMP. Its role is partly evangelical - its officers run workshops and give about forty lectures for member agencies each year in which they are urged to share intelligence via CISC, asked about their intelligence needs and offered facilities for international contacts where appropriate.

In British Columbia the equivalent provincial clearing house (CISBC) was established in May 1971, based then at RCMP Divisional headquarters, but provincial concern at an apparently rapid increase in population, levels of recorded crime and increased evidence of criminal 'organisation' was documented by a Task Force Report prepared for the British Columbia Attorney General in 1973 and led to the creation in 1974 of the Co-ordinated Law Enforcement Unit (CLEU). This built on existing experience with Joint Forces Operations and was to have an investigative division (composed of officers from the province's police forces), a legal division to advise investigators and conduct prosecutions and a policy and analysis division to focus on strategic intelligence with a view to recommending changes in statutes and resource-allocation. The second of these was not implemented because, being based on a US model, it was considered unsuitable for a system based on the traditional separation of police and prosecutors (Murray, 1982, 21-5; Walsh, 1982, 1-4). In 1995 the investigative arm was represented by two joint forces operations, one in Vancouver and one on Vancouver Island, consisting primarily of police seconded from the contributing forces.

The Policy Analysis Division developed into an umbrella intelligence unit for the province. The CLEU director hired and fired the civilian staff (57 in 1993-4) of this division who were organised for administrative reasons in various sections: resources, tactical and strategic analysis, drug strategies, open source, computer services and special projects. Their work, however, was carried out in project teams (CLEU, 1995a, 6; Engstad, 1996). CISBC was re-located to CLEU headquarters during 1994 and it appeared as though CLEU's co-ordinating role prospered.

However, within a short period the tensions within this structure had boiled over to the extent that the Provincial Attorney General appointed a review in June 1998 'to examine and assess the ability of law enforcement agencies in British Columbia to combat organized crime...' (Organized Crime Independent Review Committee, 1998, 1). The subsequent report was highly critical of the lack of coordination in the collection, analysis and dissemination of organised crime intelligence. Specifically, it noted that, whereas CLEU had originally provided a co-ordinated response, this had now been lost, the Unit was trying to do too much with too few resources, police agencies were no longer contributing officers to CLEU on an equitable basis and 'uneven CLEU leadership, a lack of accountability and workplace conflict (had) undermined morale in all ranks' (1998, vi).

In addition to the specific criticisms of the disintegration of the intelligence effort that are discussed below in chapter seven, the Policy Analysis Division came in for particular criticism. The problem, according to most of those speaking to the Review, was apparently a lack of trust that information provided to the Division would not be shared with the Ministry of the Attorney General. For example, the work of tactical analysts was highly praised but they were not taken completely on board by investigative teams because of the requirement that they report back to their managers fortnightly. Other criticisms were that policy papers had become too often designed for publication, thus limiting their utility as intelligence. The upshot of this criticism was that a majority told the Review that the Division should be split between those with operational involvement and those developing policy who should be absorbed into the ministry (1998, 36-7). There were clearly both structural and inter-personal problems at CLEU, and its disbanding immediately the Review reported was a clear recognition of the failure of the experiment that it had represented - the synergy of tactical and strategic criminal intelligence.

In any event, the resulting replacement suggested strongly that the needs of operational police had taken precedence over those for policy. In March 1999 the Attorney General announced the establishment of the Organized Crime Agency of British Columbia and that its first chief officer would be Beverley Basson, a RCMP assistant commissioner. General direction for police investigations and operations would be provided by a joint management team made up of senior police (Attorney General, 1999, March 23, 99:21). The chair of the Board of Directors, who have a similar role to a municipal police board, is a former chief constable of the Victoria, BC, police department, 4 other members are civilians and there are two police (Attorney General, 1999, April 24, 99:36).

In addition to the RCMP and provincial forces, over 100 municipal and regional police departments who have full-time CIUs are members of CISC, those with part-time units are associate members and there are 120 affiliates with investigative personnel from elsewhere in government and the private sector. One of the largest CIUs is Intelligence Services within the Detective Support Command of Metro Toronto Police (MTP). Under re-structuring proposals in February 1995, Intelligence Services was projected to include 123 police plus 24 civilians, making it the largest of the sections within the Detective Support Command. However this was fewer than the actual numbers working in this section in 1995 and smaller again than it had been some years earlier (Metro Toronto Police, 1994, 179; Sandelli, 1995).

It is a commonplace that formal structures can give only one perspective, and sometimes quite a misleading one, on the operations and interactions of organisations. In the case of CIUs this is particularly relevant, first, because of their smallness and, second, because of the nature of their work. The informal contacts that develop between people working in different agencies are frequently more important in understanding intelligence processes than official liaison arrangements. In one of the CIUs under review here, CLEU, a kind of 'formal informality' operated: in addition to the Vancouver police and RCMP, Revenue Canada, Employment and Immigration Canada and Canada Customs all assigned full-time analysts to CLEU who had direct access to their respective agency's intelligence systems (CLEU, 1995a, 2) but there were no formal Memoranda of Understanding between the agencies. Such an arrangement clearly facilitated the free flow of information but raised issues regarding accountability, which are discussed further below.

The distinctive structural interface in Canada is that between CISC and the RCMP. Given that, in Ottawa, CISC lives in close proximity to the RCMP's Criminal Intelligence Directorate, takes most of its personnel from the RCMP and lacks the independent analytical capacity that the RCMP now has, it is not surprising that outsiders see CISC as dominated by the RCMP. From an European perspective, CISC does look more like Interpol with its predominantly 'post office' function rather than the planned Europol with its intended operational capacity (see chapter two). Indeed, the only thing perhaps preventing the RCMP from formally taking over CISC is the opposition that would come from Ontario and Quebec (Levesque, 1995).

Another crucial intelligence interface in Canada is between the RCMP and the Canadian Security Intelligence Service (CSIS). The latter was created in 1984 by the separation of the Security Service from the RCMP. This came in the wake of the McDonald Inquiry into the illegal and improper behaviour of the Security Service, especially in Quebec, that recommended the creation of a separate, civilian service. McDonald's detailed examination of security intelligence issues concluded that government's need for security intelligence, often involving long-term investigations, was not best served by police officers whose priorities related to shorter-term, investigative priorities (Gill, 1994, 213-15) Even when CSIS was established, the RCMP retained responsibility for investigating security offences and the potential for turf wars was obvious; it became real within a few months when an Air India flight out of Vancouver was blown up over the Atlantic killing 329 people, most of them Canadians. Neither agency's investigation has borne fruit in terms of prosecution amid stories of fierce competition and a lack of co-operation between them. Most recently a former member of the CSIS team working on the case claimed that he had destroyed interview tapes with his informants and cut them loose rather than hand them over to the RCMP investigators because of his fear that they would be exposed (*Globe & Mail*, January 26, 2000).

The passage of time does not seem to have made it easier for the agencies to come to share information with each other - if anything, they do so less. The Security Intelligence Review Committee that oversees CSIS has referred to two main concerns. First, the 1991 Supreme Court decision in *Stinchcombe* has increased the obligation on prosecution to disclose material to the defendant in the course of prosecution and CSIS agents fear even more that their sources might be exposed in court. Second, along with their sister agencies in other countries, CSIS has become involved in the study of transnational organised crime although its contribution is supposed to be limited to the development of strategic intelligence. The Review Committee has suggested that RCMP feelings that CSIS withholds information are based on a failure to understand the distinction between tactical and strategic intelligence, implying that CSIS has none of the former that would be of use to the RCMP (SIRC, 1999, 20-24). Arguably this is rather disingenuous - while CSIS's intelligence *product* might be strategic rather than tactical, this does not mean that information it gathers cannot be of use for investigative (tactical) purposes. Indeed, in a later report the Review Committee appeared to doubt that CSIS was actually making a distinctive contribution to the investigation of organised crime rather than duplicating police activity (*National Post*, April 19, 2000).

New York

State-level agencies

The main state agency - the New York State Police (NYSP) - was discussed in chapter two. A good example of a multi-agency taskforce that operates at this level is the New York State Organized Crime Task Force (NYSOCTF). This was established by statute in 1970 and is mandated to conduct investigations and prosecutions of organised crime activities as long as they involve more than one county within New York State or cross state boundaries (Executive law, §70 - a, s. 1[a]). The Task Force is headed by a

Deputy Attorney General who is appointed jointly by the independently-elected State Attorney General and Governor. In 1996 it had nine full-time Assistant District Attorneys and 62 investigators. People currently work within one of five more-or-less permanent operational teams, one of which specialises in money-laundering. Originally these were multi-disciplinary with an accountant, analyst, attorney and investigator(s) but changes in the overall direction of the Task Force since 1994 - when there was a change in party control of the State - appear to have resulted in reduced emphasis on long term investigations and teams have become larger (NYSOCTF, 1999).

The multi-agency nature of the Task Force is manifested in a number of ways: first, personnel may be assigned from any agency such as NYPD or NYSP; second, the State Police also has their own team working in the OCTF office; third, people may be 'lent' to the Task Force for particular projects. For example, people came from the Bronx DA's office to the OCTF investigation of John Gotti that was then handed over to the federal US Attorney for Manhattan South for prosecution (NYSOCTF, 1999). Task Force priorities are established politically both in the broadest sense and occasionally more specifically. For example, in 1985 a Construction Industry Project was initiated by Governor Mario Cuomo after he had been urged by City Mayor Ed Koch to appoint a special prosecutor to investigate widespread allegations of corruption in the industry. From 1985 until 1989 the OCTF managed a major analysis and search for strategies. This involved people drawn from *inter alia* the offices of the New York County (Manhattan) District Attorney, NYSP Special Investigations Unit, NYPD, the federal Department of Labor's Office of Labor Racketeering, City Department of Investigation and the New York State Department of Taxation and Finance. A list of fifty outside scholars and experts involved is also included in the Report (NYSOCTF, 1989).

Well before the State Police was formed, in 1880 New York State appointed game protection officers whose successors now constitute the Division of Law Enforcement of the New York Department of Environmental Conservation. The Division has 297 officers of whom three-quarters are uniformed. The other main organisational unit was the Bureau of Environmental Conservation Investigations set up in 1982 that now has forty members of whom thirty are investigators working on about 400 investigations per year. At headquarters in Albany there is one investigator and an 'Analytical Lieutenant' (DLE, 1998, 5-7; Haskin, 1999). The work of the Department contains an interesting mix of 'regulatory' and 'enforcement' with the Division being the only part of the Department regularly involved with the latter. But much of the Division's work results in administrative (compliance) rather than criminal actions. NYSPIN (see chapter two) is used to access criminal histories, car and drivers licenses and to contact other agencies.   Contacts are also furthered through MAGLOCLEN and the North East Environmental Enforcement Project that is one of four regional projects jointly funded and managed by the US Environmental Protection Agency and member organisations that include five Canadian provinces. The projects facilitate information exchange and training for inspectors, investigators and prosecutors, emphasising issues such as the illegal disposal of hazardous waste.

The Division's work is essentially reactive: 'We'd like to be proactive but we're not' (Haskin, 1999). Events that are in the news or calls from the Governor's or a Senator's office will take over the agenda at short notice. Many investigations are minor but when more significant cases arise other agencies are likely to be involved. As a general rule, the more resources that are required by an investigation, the more likely it is that some formal MOU will have to be negotiated by the agencies concerned in order to meet data exchange laws, the distribution of costs and, if relevant, determine the allocation of any 'spoils' that might eventually result from civil or criminal forfeiture. One such involved an investigator going undercover for 18 months as part of an FBI investigation into corruption between municipalities and a solid waste disposal contractor. Needless to say, investigators prefer to deal with contacts informally in other agencies if they can do so to avoid the inconvenience of managers being involved in the negotiation of MOU that slows down the process for everyone.

Policing the City

NYPD is not only the largest but, after the FBI, probably also the best-known US police agency thanks to many feature films, extensive research, regular corruption scandals, and the TV series *NYPD Blue*. Major innovations in recent years have to a greater or lesser extent been imitated in many other forces and are indicative of current debates as to policing strategies. In 1984 the Community Patrol Officer Program was launched in which officers in pilot precincts were relieved from responsibility for reacting to calls in order to identify and plan solutions to the area's problems in an early manifestation of what became 'problem-oriented policing' (Goldstein, 1990, 58). Some early assessments of the pilot projects were quite optimistic and, because of rising fear of crime, the system was extended city-wide in 1990 but with only limited impact (Silverman, 1999, 54-63). In 1993, however, newly-elected Republican Mayor Rudolph Guiliani announced that he believed community policing to be too soft and under his new police commissioner William Bratton, 'zero-tolerance' policing was launched amidst a re-structuring of NYPD that sought to empower precinct commanders (Silverman, 1999, 79-96).

Zero-tolerance policing is based on James Q. Wilson's 'broken windows theory' that ignoring minor crime contributes to a decline in the quality of life and encourages major crime (Wilson & Kelling, 1982). Bratton had earlier been head of one of New York's transit police departments where he had started more aggressive policing of 'fare beating', begging and graffiti in the subways. Extending these policies to the streets and parks of the city certainly continued the steady fall in crime rates that began in 1990, including murder rates reaching their lowest levels for thirty years. In 1990 there were 2300 murders in the City - this fell steadily to 800 in 1997 (Karmen, 1998, 37). But there has been considerable controversy over the extent to which this was caused by the change in strategy. Some criminologists agree with Bratton himself that policing policy made the difference (Bratton & Knoblach, 1998) while others disagree, pointing out that other cities with different policies saw similar changes (Walker, 1999, 171-3). Zero-tolerance policing was also held responsible for a significant increase in citizen complaints against the police - between January 1993 and mid-1996 the monthly average number of complaints against police recorded by the Civilian Complaints Review Board increased from 250 to over 500 and for the next two years was between 400 and 450 (CCRB, 1998, 63). Further, as always with police, one needs to be sceptical regarding the statistics - murder figures are relatively difficult to massage but the greater accountability pressure on precincts and departments that resulted from Compstat (discussed in some detail in Silverman, 1999, 97-124) did lead to manipulation and some forced resignations (*New York Times*, Feb 28, 1998, B1). Other major examples of police abuse of power showed that NYPD had not lost its ability to generate great pubic controversy, for example, the torture of Abner Louima, a Haitian immigrant in a precinct station in 1997 (of which officers were later convicted) and the shooting of Amadou Diallou in February 1999 (of which four officers were later acquitted).

The more specific police intelligence function in NYPD has been no more immune from periodic controversies. A force this size and with the consequent degree of specialisation can clearly sustain specialist intelligence functions in a number of areas but there are three in which there is a specialist named Intelligence Unit - and the location of these provides a hint of the ambiguous position of intelligence within policing, both for police and public. The Intelligence Division exists separately from the main operational departments and precincts and is the historical heir of the 'Red Squad' - these existed in most large urban forces after the 'red scare' of 1919-20. Starting life in New York as the 'Bomb Squad' in 1919 it later became the Radical Bureau and compiled files on members and supporters of the Communist Party and strikers, exchanging information both with federal and other police agencies. After Fiorello LaGuardia became Mayor in 1934 surveillance activities continued but disruption of outdoor meetings and strikes was reduced (Donner, 1990, 40-48).

The Red Squad was officially the Bureau of Special Services (BOSS) during the 60s when it enjoyed a much greater degree of autonomy than most of its equivalents elsewhere (Donner, 1990, 155). However, in the 1970s real political and legal pressure was applied to BOSS operations that, slowly but steadily, eroded its autonomy. In May 1971 a class action lawsuit was filed against the Bureau for a range of abuses of First Amendment rights including the planting of informants, the provocation and incitement of criminal acts, intimidating surveillance and confrontation and wiretapping. Lack of funding slowed the pursuit of the case - *Handschu* - and the 1980 settlement was only approved by the court in 1985 (Donner, 1990, 356-7). In 1973 new intelligence guidelines linked surveillance explicitly to 'probable cause' of a criminal act and a purge of files were announced. The names index was reduced from 1.22m to 240k people and from 125k to 25k organisations. Files on individuals were reduced from 3,500 to 2,500 and for organisations from 1,500 to 200 (Donner, 1990, 158). But guidelines are just that - they do not necessarily produce reform; for example, the failure of the police to follow the new guidelines led to the dismissal of a case against a black activist in 1975 (Donner, 1990, 191).

Since a force reorganisation in July 1998, the head of Intelligence Division is responsible directly to the Commissioner who, consequently, has a significant say in the Division's priorities. The Division is now divided into three sections: municipal security, public security and criminal intelligence, each with about one third of the overall 340 people employed there. Municipal Security provides protection for the mayor and other elected officials and includes both a uniformed section for patrolling key locations such as City Hall and also a plain clothes detective squad for personal protection.

Public Security is itself in three sections: first, a threat assessment unit to examine the 1300 or so threats to public officials and buildings received each year. Second, a 'Special Services Unit' looks at political groups who are engaged in 'First Amendment activity' that is constitutionally protected but which might constitute a threat of criminal activity and works under the *Handschu* Directive. In 1989 a federal court held that the police had broken the *Handschu* agreement as to acceptable surveillance practices when it monitored and taped the views expressed on a black-oriented radio station. But allegations that the Intelligence Division maintained a 'black desk' to compile information on black political figures were rejected (*New York Times*, October 10, 1987, B35; July 22, 1989, B29). The third section has various functions, including providing protective security for VIP visitors to the City that involves co-ordinating with other agencies involved, for example, the Secret Service. Also, this section has a general liaison function with other forces and countries.

When Howard Safir became Commissioner in 1996 most intelligence work was being carried out in the Organized Crime Control Bureau and, to a lesser extent, the Detective Bureau. Given his background in federal enforcement, he sought to reinvigorate criminal intelligence and re-created what is now the Division's criminal intelligence section. This runs a 24-hour intelligence centre in partnership with the New York/New Jersey HIDTA - 95% of the staff are NYPD analysts who seek to provide a 'one-stop shop' for other agencies. There is a Gang Intelligence Unit and an Analysis Unit that includes a small 'cell' created with four detectives for organised crime issues. In attempting to prevent turf disputes between the Intelligence Division and elsewhere, the Commissioner laid down that it was not to 'make cases' and, if it does, then they will be handed over to the most appropriate department. Some 90% of the product of the Division is said to be 'strategic'; 'tactical' intelligence regarding organised crime will be the prime goal of the Organized Crime Control Bureau (Oates, 1999).

Organised crime is the second area in which NYPD has a specific Intelligence and Analysis Unit. Interwar enforcement against organised crime, such as it was, was often federal in origin because of the unwillingness and/or inability of local police to act. For example, the US

Treasury had primary responsibility for the enforcement of Prohibition but, in many areas of crime, federal agencies lacked jurisdiction. For a number of reasons - the socio-economic structure of New York, the increasing political salience of the 'Rackets' in the interwar period and the election or appointment of such reformers as LaGuardia and Thomas Dewey, police in New York were brought up against the problems of enforcement in this area earlier than most local forces.

It was only in the 1950s that more forces became involved after the first Congressional hearings into 'organised crime' with Tennessee Senator Estes Kefauver in the Chair. The televised hearings brought the issue before the public for the first time even though the final report can be criticised, in part because of its acceptance of the crude alien conspiracy theory of organised crime that suited well the McCarthyite atmosphere of the times (Abadinsky, 1997, 503-07). Yet criminal intelligence units were established in a number of forces as a result, a development given further impetus by the recommendation of the President's Commission on Law Enforcement and the Administration of Justice in 1965. A further boost was given to the development of specialist intelligence mechanisms by the recommendation of the Kerner Commission into the ghetto rebellions of the mid-1960s (Donner, 1990, 77). Given the similarity of the dominant official 'explanations' for these problems - conspiracies of predominantly ethnic minorities - it is not surprising that in some areas political intelligence was simply assimilated into organised crime units, but in some areas such as New York City the two were kept separate (Donner, 1990, 67).

Yet the ability of NYPD to deal with organised crime has always been mediated by corruption. Since its establishment in 1844 the highly decentralised force had been an important part of local power networks of politicians, capital and labour in which favours and pay-offs oiled the wheels of 'business'. NYPD 'was more or less a branch of Tammany' until the first World War according to Abadinsky (1997, 85) or until LaGuardia's election, according to Woodiwiss (1988, 54). In any event, since the late nineteenth century there has been a twenty year cycle of corruption scandals followed by special commissions of investigation (Walker, 1999, 254-5). Following the Knapp Commission report in 1973 Commissioner Patrick Murphy both increased the size of the then Internal Affairs Division and decentralised to a network of Field Internal Affairs Units. However, an investigative structure requires the will to operate it and the Mollen Commission reported in 1994 that Internal Affairs at headquarters had been doing little. Between 1988-92 its ninety investigators shared out an average of 133 cases per year between them - many of them trivial - while devolving to the 270 investigators in the field units over 2,500 cases per year (Mollen, 1994, 85-90). The Mollen Commission recommended *inter alia* a range of improved intelligence techniques including debriefing informants, profiling and a more proactive approach (1994, 138-142; see also Silverman, 1999, 49-54). The Internal Affairs Bureau now has a specialist intelligence section but it remains to be seen whether this will break the cycle.

Policing the Port

New York's commercial pre-eminence is largely due to the historical significance of its deep-water harbour where the nature of the labour process - erratic and inelastic demand for labour to unload and load at short notice - generated extraordinary profits and was highly attractive to criminal enterprises. The apotheosis of organised crime control of the waterfront was reached when Joseph Ryan was President of the International Long-shoremen's Association (ILA) between 1927 and 1953 (Abadinsky, 1997, 405-6). The ILA had effectively a closed shop agreement with the shipping and stevedoring companies but the workforce was unionised in name only:

> The ILA was, in effect, a company union run by gangsters. It made no pretence of maintaining a defence fund because its function was to prevent the men organizing and to break strikes. The union gangsters only drew a nominal salary from the treasury - they were regularly paid off by the companies and allowed to operate any pierside rackets they chose. (Woodiwiss, 1988, 63)

Wildcat strikes were the only weapon that longshoremen had to try to improve conditions and the Union 'goon squads' were used to violently suppress rank-and-file activity on the pretext that it was communist-led. This unholy brew of ideology, violence and profit led to a complicity of vested interests until 1951 when State Governor Thomas Dewey ordered the State Crime Commission to investigate waterfront conditions. In 1953 both federal and state legislation was passed approving the creation of the bi-State Waterfront Commission.

The Waterfront Commission of New York Harbor provides an interesting example of a 'special district' cross-state agency that has both regulatory and enforcement powers. Its mission is:

> to investigate, deter, combat, and remedy criminal activity and influence in the Port of New York-New Jersey and to ensure fair hiring and employment practices, so the Port and region can both grow and prosper. (Waterfront Commission, 1997, 2)

The Governor of each state appoints a Commissioner for three year terms, often extended. An executive director is responsible for day-to-day operations of the six divisions into which the Commission is divided. The regulatory functions are extensive, for example, licensing the stevedore companies and screening and registering the longshoremen, checkers, hiring agents and pier guards. Since 1953 the Commission has controlled the Register of those entitled to seek work and is empowered to control its size by removing workers who do not make themselves available for work regularly. An agreement negotiated at the time of containerisation gave registered workers a guaranteed annual wage but, in order to control the number of guaranteed payments, since 1966 the Commission has been further empowered to control the entry of new workers into the industry (Waterfront Commission, 1997, 4).

The Police Division is primarily investigative - at some time in the past they had uniformed patrol police but did not find it productive. The Commission Police carry out investigations of both criminal activity and violations of the labour-management compact, and conducts background checks on applicants for licenses. The investigative priorities of the Commission are established almost entirely reactively, as 'tips' from a worker-cum-informant or some other local, state or federal agency. As with the OCTF, the law governing the Commission requires them to share information with other agencies but, again, the degree of actual sharing and co-operation with other agencies is less a matter of formal agreements than dealing with people who are trusted (McGowan, 1999).

Clearly, conditions on the waterfront have changed significantly in the last fifty years but changes in the labour process have probably contributed as much to this as the actions of the Commission. Initially, the Commission Register became the basis of a blacklist of union militants and suspected communists that did legally what the gangsters had been doing for years (Woodiwiss, 1988, 65). Eventually the Commission eliminated the 'shape-up' (daily recruitment of casual labour) and 'public loading' rackets but the dramatic collapse in the numbers and nature of employment associated with containerisation was also a major factor. After the first decasualisation in 1955 there were 31,574 left on the Register, by 1969 there were 20,627; 1979: 10, 956; 1989: 5,846; 1997: 2,848 (Waterfront Commission, 1997, 15). The ILA shifted the focus of its continued domination of the port from hiring rackets to reducing competition between stevedores and other contractors who paid 'rents' in order to secure work; otherwise it moved operations to the booming - and unregulated - port of Miami (Abadinsky, 1997, 407).

Patrolling in the port area is carried out by the Port Authority of New York and New Jersey Police. Formed in 1928 and now consisting of over 1,400 officers its jurisdiction covers the Port District in both states and includes law enforcement and fire fighting. Its specific responsibilities are the airports, bridges, tunnels, bus and port terminals as well as the World Trade Center and PATH transit system between Manhattan and Newark. Since aspects of their substantive and geographical mandates are identical, for example cargo theft, it is perhaps curious that the publicity materials prepared by both Commission and Port Authority Police refer to their co-operation with numerous local, state and federal agencies but not with each other.

Transit Police

The specialised transit police provide another interesting example of the fragmented nature of policing in New York. As well as the Port Authority police taking responsibility for their own PATH transit system, Amtrak, that provides the main long-distance rail routes South to Washington and North to Canada and New England, has its own police department. The Metropolitan Transportation Authority (MTA) also has its own police department but it does not have jurisdiction over all the routes that the Authority has. The MTA Police Department has itself just come about as a result of a merger between the two separate departments that covered respectively the Northern line from Grand Central up-state and to Connecticut and the Eastern line from Penn Station to Long Island. What had been another separate transportation police department covering the New York subway was integrated within NYPD in 1995. Transit policing is much more about patrol than investigation, for example, fewer than seven percent of MTA police are investigators, roughly half the usual proportion in police agencies.

Particular locations require more or less formal agreements regarding jurisdiction, for example, Penn Station is the responsibility of the MTA police but is also used by Amtrak and connected to the subway. Grand Central station is a 'Business Improvement District', and there is public-private collaboration in the policing of the complex that includes both the Metropolitan Life building and the station. The routes themselves will also pass through many different jurisdictions, for example, the MTA's Long Island line passes through 35 different police areas in which the MTA police have joint jurisdiction for offences occurring on railroad property.

A problem facing the transit police is that the crime with which they deal - mainly 'quality-of-life' and auto-theft from station car parks - figures relatively low down the priority scale of the larger police agencies with whom they need to co-operate and they may not even receive the relevant information despite formal agreements that they should. The introduction of Compstat in NYPD has led to some slight increase in information sharing between agencies, for example, the MTA police will be invited into Compstat meetings if relevant. Otherwise, the main arena of potential co-operation is in emergency planning where transit police will be drawn in to networks such as the City wide Terrorist Task Force Group and the Mayor's Office of Emergency Management that deals with VIP visits to the City (Loverso, 1999).

Investigatory Commissions and Departments

The State Crime Commission established by Governor Thomas Dewey in 1951 led not only to the Waterfront Commission but also provided much evidence of the corruption of law enforcement involving sheriffs, police departments and district attorneys and recommended the establishment of a permanent Commission of Investigation. This now has a broad mandate 'to investigate any matter concerning the public peace, public safety and public justice'. It will examine fraud, corruption and mismanagement in general but also has a specific responsibility to investigate cases where organised crime is alleged to be corrupting the enforcement of state law. The Commission is investigative only - evidence of criminal behaviour will be referred to a district or US Attorney. It has subpoena powers and may grant immunity to witnesses. As well as making specific reports to prosecutors where appropriate, the Commission also has the authority to conduct public hearings and issue public reports and to make proposals to the Governor and state legislature for remedial legislation (www.sic.state.ny.us June 1998).

In the city the struggle against political corruption has been waged officially for longer. The Office of the Commissioners of Accounts was established in 1873 as a reaction to the excesses of 'Boss' Tweed and his Tammany Hall colleagues - the *New York Times* estimated that between 1869-72 they extorted or stole $200 million (Winslow & Burke, 1993, 1). Since then, the history of what is now known as the Department of Investigation (DOI) has ebbed and flowed along with the broader tides of reform and corruption in the City and in recent years its investigations have been significantly entwined with the political fate of several City mayors.

Shortly after his election to a third term in 1985 Ed Koch's administration was hit by the first of a series of scandals that were eventually seen as responsible for his failure to win a fourth term in 1989. It involved bribery and fraud in the purchase of proposed hand-held computer devices for issuing summonses by the Parking Violations Bureau and the DOI itself was criticised for a failure to investigate earlier allegations against the Bureau. The DOI Commissioner himself had resigned under pressure in January 1986 after allegations that he had mis-used DOI resources. Later the same year the State Commission of Investigation criticised the integrity of one of the City's inspectors general in a case involving the Taxi and Limousine Commissioner, who was later convicted. As a result of these pressures, Koch integrated the inspectors general (who had worked within the departments they were monitoring) into the structure of the DOI. This increased the staff of the agency to almost 700 and, as importantly, made inspectors general responsible solely for the investigation of criminal matters, relieving them of their role in staff discipline. However, further scandals contributed to Koch's replacement by David Dinkins, the first African-American to be City Mayor (Winslow & Burke, 1993, 71-7).

Dinkins appointed Susan Shepard, who had worked at the State Commission, to head the DOI that then embarked upon a period of what appeared to be unusual independence, culminating, ironically, in the demise of Dinkins himself. Under Shepard the 21 inspectors general were 'clustered' into five specialised groups to achieve a greater consolidation of investigative resources at a time when they were significantly reduced because of the City's financial crisis. By 1992 staffing was down to 353 (Winslow & Burke, 1993, 83-4; DOI, 1993, p.5).   Yet when the

Transportation Commissioner complained to DOI of the influence being peddled on behalf of Lockheed (owners of Datacom who had been at the centre of the earlier Parking Bureau scandal) to obtain the City contract for the privatisation of the Bureau, the DOI this time pursued the investigation vigorously and, in its report published three months before the 1993 election, implicated a number of senior officials in Dinkins' administration. The Mayor, having achieved office in part because of the first Parking Bureau scandal, lost it because of another when he was defeated by Giuliani in the election (Benjaminson, 1997, 155-65).

Shepard was replaced by Howard Wilson, a friend and associate of Giuliani's who had worked in his campaign, and who, unlike Shepard, attended the Mayor's daily cabinet meetings. A willingness to investigate even the allies of the Mayor was apparently replaced by a tendency to investigate the Mayor's opponents (Benjaminson, 1997, 170-81). Now, the DOI's staff is about 350 including attorneys, accountants, civilian investigators and 35 NYPD detectives (Burke, 1999). The fifteen inspectors general specialise in one or more city agency, still clustered into five groups: capital construction, correctional services, licensing and inspection, procurement, real property and finance, and public assistance and contracting/grant agencies. Of the complaints received by DOI 60% or more will be referred on to other agencies and about 10% will result in new investigations (860 out of 8716 in 1997/8).

Although the jurisdiction of the DOI is over 'mayoral' agencies, therefore formally including NYPD, it does not investigate the police. In the past the DOI did investigate allegations of police corruption, for example, in 1969 then Commissioner Ruskin reported that there had been seventeen investigations but, to judge from the rising tide of news stories, this was doing little but scratch the surface of the problem. A long and detailed story in the *New York Times* in April 1970 set out the systematic nature of the corruption and Mayor John Lindsay appointed what became known as the Knapp Commission. DOI kept up individual investigations of cases while the Commission sat but the Knapp Commission was critical of DOI efforts. In its final report it criticised Arnold Fraiman, Commissioner between 1966-68 and then a State Supreme Court Justice, concluding that he:

> ...failed to take action that was clearly called for in a situation that seemed to involve one of the most serious kinds of corruption ever

to come to the attention of his office, and which seemed to be precisely the sort of case his office was set up to handle. (quoted in Winslow & Burke, 1993, 57-8)

In 1977 the former head of the DOI police unit was convicted of lying about a 1966 bribe offer and of warning police officer Frank Serpico to stop pursuing corruption allegations (that later provided a significant part of Knapp's agenda) or he risked winding up 'in the East River, face down.' (quoted in Winslow & Burke, 58 fn115) In the light of the Knapp Commission's criticisms of DOI, an Office of Special State Prosecutor was created with city-wide jurisdiction over allegations of corruption within the criminal justice system including the police and the DOI left the field. Given the steady degeneration of NYPD's own internal affairs division, especially after 1986, as documented by the Mollen Commission (1994, 70-89), however, it is clear that the problem of establishing some effective combination of internal and external oversight of the police remained. Now, DOI only investigate the NYPD obliquely, for example, it is currently looking at 115 cases in which the Civilian Complaints Review Board sustained complaints against the police but in which the cases then 'disappeared' before the Police Department took any action (Burke, 1999).

Another example of an investigatory commission is the New York State Ethics Commission. This was established by the Public Officers Law 1987 that resulted from the then perception of widespread corruption including the first Parking Violations Bureau scandal in the City. Separate commissions were established for each branch of government, that for the executive branch has five members appointed by the Governor, one each being nominated by the Attorney General and Comptroller.   The Commission has a staff of five lawyers, three investigators and support staff and meets about every six weeks to make its decision. The main areas covered by the law are: the need for 'policymakers' to make financial disclosures, conflicts of interest and certain outside activities, gifts and honoraria, and post-employment restrictions. The Commission distributes, collects and audits financial disclosures - of the 17,000 it receives annually it actually audits about 1,000. It issues advisory opinions when requested by a state employee - most are in the form of non-binding advice but it also issues about thirty each year that are formal and binding.

The investigative function relates to alleged violations of the Public Officers Law and the Commission receives about 150 complaints each year many of which are re-directed since they are not within the mandate. The Commission has subpoena powers but the Commissioners prefer to pursue public education and restitution rather than project the image of a law enforcement agency. As such their role can be contrasted with that of the state inspectors general who were set up in 1986 and have cultivated a more aggressive image. If the Commission finds that a minor violation has occurred then their role is to notify the person and publish the decision. It is the job of the agency concerned to decide whether disciplinary action should follow. Only in the most serious cases will the Commission levy a civil penalty or refer the violation to a prosecutor; in return cases are sometimes sent from other police or governmental agencies where a criminal case cannot be sustained and the Commission is offered the chance to investigate (Campbell, 1999).

The role of prosecutors

Something that clearly distinguishes the US criminal justice process from that in either Britain or Canada is the more active role of prosecutors in investigations. The emphasis for prosecutors is clearly upon bringing cases to court rather than developing intelligence but the long term nature of some of the investigations undertaken by some prosecutors certainly involves the development of tactical intelligence. Inevitably, the fragmentation of US government is reflected in the structure of prosecutors - each of New York State's 62 counties has a District Attorney, elected every four years, of whom 51 work full time and 11 part time. 2,662 full-time assistant district attorneys (ADAs) were employed in the state in 1996, 68% of them in the five counties that constitute New York City (DCJS, 1996).

The largest office is that in New York County that covers Manhattan: it has over 1200 staff of whom 500 are ADAs. The office is divided into four parts: Trial Division, Investigations Division, Appeals Bureau and Narcotics Bureau. The last of these is amalgamated into the Office of Special Narcotics Prosecutor (see below). The Trial Division has about 300 lawyers who are responsible for the prosecution of street crime and the Division includes a number of smaller specialist bureaux. For example, the Homicide Investigation Bureau was established in 1983 to seek greater co-ordination of law enforcement after the crack epidemic and associated increase in gang murders had hit Manhattan. The Bureau has ten

ADAs, ten former NYPD detectives as investigators and five analysts, each of the ADAs specialising in a specific area (Arsenault, 1999).

The Investigation Division concentrates on long-term investigations and while only 5% of the lawyers are in this Division the work of the 100 civilian investigators and 50 NYPD detectives on permanent assignment is overwhelmingly concentrated here. There are a number of specialist sections, including, for example, the Rackets Bureau that has thirty lawyers including ten in a Labor Racketeering Unit.

In a recognition of the problems being generated by 'cross-border' crime, in 1971 the State government created the Office of the Special Narcotics Prosecutor (OSNP). In 1997 the OSNP had 180 staff, 87 of them lawyers and twenty investigators (OSNP AR 1997). Over three-quarters of the lawyers are from the Manhattan District Attorney's office because of the amalgamation of its own Narcotics Bureau into the OSNP. The Special Prosecutor herself is appointed by the five county District Attorneys and the office is divided into two main sections. The Trial Division prosecutes felony drug offences at the 'street-crime' level involving three main techniques of buy-busts, surveillance of sales and employing search warrants. The Special Investigations Bureau undertakes longer-term and more complex investigations that, given New York's size and significance as a port almost inevitably involve collaboration with other local, state and federal agencies, in particular NYPD's Narcotics Division, the New York Drug Enforcement Task Force and the DEA's New York Region Office. In addition to the OSNP's own eleven investigators, nine of them former NYPD detectives, there is a further team of seven detectives assigned from NYPD's Narcotics Division who provide primarily an array of technical support for surveillance (OSNP, 1997).

Given the preponderance of lawyers over investigators in the OSNP it is not surprising that the office's priorities are established almost entirely reactively to what the police bring them, especially in the Trial Division. It is impossible to identify any simple method by which priorities are established but, with DAs being elected, clearly politics is significant. Also, it is clear that different investigations might well derive from different analyses of how drugs markets operate and, of course, much will depend on the negotiation with other agencies in selecting investigations that are likely to be fruitful. For example, in 1997 the Trial Division worked with the NYPD's Northern Manhattan Initiative that sought to drive drug dealers out of Washington Heights as part of a broader 'zero tolerance' policy (e.g.

Silverman, 1999, 197-99). As such this led to the targeting of quite small drugs gangs and can be contrasted with the strategy of targeting-up that seems to be preferred by the Special Investigations Bureau. On occasions, of course, a policy of short term arrests and/or disruption may contribute to a longer-term investigation but sometimes the two strategies may conflict.

Another significant dimension facing local prosecutors is the presence of federal US Attorneys, for example, in New York City there are two - one for the Southern and one for the Eastern District. Over the years there have been some spectacular problems, for example, when the Southern District US Attorney indicted someone who was acting as an informant for the Manhattan District Attorney in the context of more general obstruction of his BCCI investigation by the Justice Department (*The Guardian*, October 2, 1992, 14). The BCCI affair is, of course, an extraordinary example of the extent to which the relationships between, on the one hand, licit and illicit organisations and on the other, law enforcement and intelligence agencies may be interwoven (e.g. Passas, 1993). Several interviewees commented on the many occasions on which there had been serious feuding over cases despite the efforts made in negotiations between OSNP and the federal US attorneys to protect everybody's interests. Clearly this will not always be possible in a structure so permeated by electoral and organisational politics and where there are a number of strategic issues to be determined.

For example, the substantive criminal law in the US is state-made but in recent years there has been a growing tendency for the 'federalisation' of offences. This has occurred, for example, following specific 'moral panics' or in the wake of real or imagined failings of state agencies to enforce adequately state law. Consequently there may be room for strategic choices to be made as between different legal regimes and this can be a factor in encouraging the co-operation of agencies at different levels of government. So, for example, the Suffolk County District Attorney (on Long Island) will work happily with the Division of Law Enforcement of the State Department of Environmental Conservation since crimes may carry greater punishment if they are dealt with under state law rather than county ordinance (Haskins, 1999). Or, at a 'higher' level, if a RICO investigation is being undertaken then the involvement of a federal agency may well suit a state or local agency because the federal RICO statute is perceived as more prosecutor-friendly than the New York State equivalent (Arsenault, 1999).

'The Feds'

In 1996 there were 74,500 full-time federal law enforcement officers in the US authorised to make arrests and carry firearms. The main roles performed by these officers are criminal investigation (43.0%), non-criminal investigation (12.9%), patrol (16.0%) and corrections (21.2%). 14% were women and 28% members of a racial or ethnic minority (www.usdoj. gov/bjs). Although they are spread across many agencies, most officers are concentrated in a few large agencies, mainly within the Justice and Treasury departments. The size of the largest agencies is below.

Table 5.1 Main US federal law enforcement agencies, 1996

Agency	Officers, June 1996	%age change 1993-96
Immigration and Naturalization Service (INS)	12,403	+31
Federal Bureau of Prisons	11,329	+13
Federal Bureau of Investigation (FBI)	10,389	+3
US Customs Service	9,749	-4
Internal Revenue Service (IRS)	3,784	+5
US Postal Inspection Service	3,576	0
US Secret Service	3,185	na
Drug Enforcement Administration (DEA)	2,946	+5
Administrative Office of the US Courts	2,777	na
US Marshals Service	2,650	+23
National Park Service	2,148	0
Bureau of Alcohol, Tobacco and Firearms (BATF)	1,869	-5
US Capitol Police	1,031	-5
US Fish and Wildlife Service	869	+40

Source: US Department of Justice, Bureau of Justice Statistics (1998), www.ojp.usdoj.gov/bjs/ May 1999

No attempt is made here to describe all of these larger agencies, let alone the 113 identified by Bayley (1985). Attention is given to those agencies most centrally involved in national law enforcement and who therefore are most likely to be involved in the networks of state and local agencies particularly when issues of federal law, cross-state and transnational investigations arise.

Department of Justice

The Immigration and Naturalization Service (INS) has been within Justice since 1940 - it was moved there from the Department of Labor because of increasing international tensions - and is responsible for enforcing the law relating to the admission of aliens and naturalisation procedures. Its two main programmes are correspondingly enforcement and examinations. The former includes the uniformed border patrols, plainclothes investigators (working, for example, in the areas of alien-smuggling), intelligence (in support of both examinations and enforcement) and detention and deportation (including the maintenance of detention facilities for those awaiting deportation. Since 1988 - when the Anti-drug Abuse Act was passed establishing the Office of National Drug Control Policy, the so-called Drug Tsar - the INS has been drawn into more drugs enforcement both at the border and within the country (www.usdoj.gov/ins May 1999).

The FBI started life within Justice as the Bureau of Investigation in 1908 and is by far the best known of the federal law enforcement agencies, not least because of its successful self-promotion during the incumbency of J. Edgar Hoover. Taking over the small and scandal-ridden Bureau in 1924, Hoover sought to re-make it as a technologically-advanced investigative agency free of corruption and by the second world war he had largely succeeded. Hoover's strict personal control of the Bureau ensured that, certainly by comparison with many local police agencies, its agents were honest but there were already signs of some of those traits that would later see Hoover lead the Bureau down several blind alleys. For example, his obsession with favourable publicity led to distortion both of the crime threats to the US and the Bureau's effectiveness. In the 1930s the Bureau's reputation was constructed on the pursuit of bank robbers such as Dillinger rather than the organisationally more sophisticated criminal organisations prospering during Prohibition. While it can be argued that the FBI enjoyed legal jurisdiction over the former but not the latter (e.g. Poveda, 1990, 112), it is also the case that Hoover preferred to keep his agents away from areas in which policing was more prone to corruption and in which it was harder to produce statistics of success.

During the second world war the FBI concentrated on internal security matters and, with the almost immediate onset of the Cold War, continued them in the anticommunist cause. Hoover's own beliefs, and the increasing significance of what became known as 'McCarthyism' provided the Bureau with a secure bureaucratic base upon which to construct its role as 'defender of dogma' (Edelman, 1964, 69-72) in the postwar period. As far as 'crime' was concerned, Hoover was able to maintain his position that the Mafia was a myth and although he commissioned Bureau research on the issue of organised crime after the Appalachin convention was uncovered in New York in 1957, it was not until further pressure from Attorney General Robert Kennedy and then Senator McClellan's hearings in 1963-64 that Hoover would acknowledge publicly the scale of the problem (Poveda, 1990, 115-6).

However, at no point would the investigative resources being put into organised crime by the FBI be more than minimal while Hoover was alive. One reason for this was Hoover's reluctance to co-operate with other federal agencies unless the FBI was clearly in charge. During the 1960s the main federal law enforcement effort against organised crime developed in the context of multi-agency taskforces and by 1971 they were operating in 17 cities but the FBI had contributed only four agents compared with the 224 from other federal agencies (Poveda, 1990, 117; Goldfarb, 1995, 45-50). Throughout this period the Bureau still targeted the Antiwar and Black civil rights movements far more than crime as such but, when Hoover died in 1972, the public reaction against the 'counter-subversion' programmes that had dominated the FBI's work had already begun and, soon after Hoover's death, the number of political intelligence investigations conducted fell away dramatically, even before the Church Committee hearings of 1975-6 (Poveda, 1990, 79-80).

In the wake of those hearings the investigative resources available to the FBI actually declined for several years and only began to pick up again after 1980 when the Bureau benefited from the resurgence of security concerns that had brought about the election of Ronald Reagan. By this time the Bureau had started to re-orient itself from the Hoover days both in terms of targets and methods. The discredited internal security investigations had largely been replaced by concerns with white-collar crime, political corruption and organised crime, though not entirely as the CISPES investigation showed (Stern 1988). Regarding methods, whereas Hoover had discouraged proactive investigations of 'crime' although not of politics, under directors Kelley and then Webster, the Bureau was to start making greater use of proactive, undercover measures (Poveda, 1990, 118-21).

The Drug Enforcement Agency (DEA) emerged as such in 1973; it was originally established in 1930 as the Federal Bureau of Narcotics - an independent agency within the Treasury - with Harry J. Anslinger as its Commissioner who then served until 1962. It is not only in terms of the length of service that Anslinger can be compared to Hoover. He was also a successful publicist for FBN views of the world and analyses of the nature of the drug problem in the US: essentially, it was a police rather than medical problem and arose mainly because of foreign and/or ethnic conspiracies to supply drugs to unwitting Americans. The answer was therefore to stop supplies of drugs entering the US (Woodiwiss, 1993, 1-14). During Prohibition most criminal organisations involved in the manufacture of illegal liquor were unambiguously American but once that was repealed there was no more challenge to Anslinger's view than there was to Hoover's regarding the threat of American communism. As such the careers and fortunes of the two men ran in parallel and yet there was also fierce bureaucratic rivalry between them. While Hoover concurred on the serious-ness of the alien threat to the US, for many years he disagreed with Anslinger's identification of the Mafia as the major source of crime in America, there was constant rivalry over budgets (Poveda, 1990, 114) and, Anslinger's definition of the drug problem put him ahead of the game in 'internationalising' US law enforcement abroad (Nadelmann, 1993, 103-07).

The increase in drug use in the US in the 1960s led to a series of legislative and organisational shifts that marked the first 'war on drugs' declared by Nixon and subsequently renewed by Reagan and Bush. The FBN itself was virtually abolished in 1968 when massive corruption emerged - about a third of its personnel, almost the entire New York office, was fired, forced to resign, transferred or convicted (Woodiwiss, 1993, 12-3). Several agencies were merged and placed within Justice as the Bureau of Narcotics and Dangerous Drugs; the DEA emerged a few years later after a series of turf battles (Nadelmann, 1993, 140). Half of its 7,500 staff are special agents and it also has 420 intelligence specialists. Now, almost 500 staff work abroad (www.usdoj.gov/dea April 1999) where their approach is relatively activist; until the FBI's announcement of an operational office in

Budapest, DEA agents were the only ones actually co-operating operationally with foreign agencies rather than as liaison.

The DEA retains a view of the nature of the drugs and crime problem that is essentially unaltered from that it presented to the Kefauver hearings nearly 50 years ago. As demonstrated in a speech to an international conference in March 1999 by the DEA head, Tom Constantine, the drugs problem is still analysed in purely 'supply-side' terms, the only shifts have been in the identity of the syndicates controlling the supply. In the 1950s and 1960s it was the US-based Mafia of Italian-Americans controlling heroin supply; once the French Connection was broken they were replaced by Colombian 'cartels' supplying cocaine and, with the eclipse of the Medellin and Cali cartels, they have been replaced in the 1990s by Mexicans. But the organisational model remains hierarchical, for example: 'The Cali organizational structure was similar to American organized crime in its tight control of workers and compartmentalized hierarchy...' So, DEA strategy has remained directed at the 'kingpins':

> DEA's approach is threefold. First, attack the principal leadership of these international organized crime syndicates who operate outside of our geographical boundaries by building solid cases against them and indicting them, often repetitively, in U.S. jurisdictions. Second, attack the surrogates of these international drug lords who operate on U.S. soil, represent the highest levels of the command and control structure of these organizations and are responsible for carrying out the orders of their bosses. And third, attack the leaders of the domestic gangs who distribute drugs in local communities and are responsible for the vast majority of the violent crimes that are associated with their drug activities. (Constantine, 1999)

Treasury Department

Having been established in 1789 US Customs is the oldest federal law enforcement agency in the US but its 8,000 miles or so of land borders in the North and the Southwest have always presented a significant challenge to enforcement (Nadelmann, 1993, 22-31). Customs' mission includes both enforcement and revenue-raising. The nature of its task has always placed a premium on inter-agency work and current information provided by the

Service makes reference to a number of initiatives with which it co-operates including FinCEN and HIDTAs (discussed in chapter two). What is surprising, given the current significance of drug interdiction for US agencies, is that co-operation with the DEA is not mentioned in this material. For example, it refers to the Border Coordination Initiative as a management strategy negotiated between Customs and INS 'to more efficiently interdict drugs, illegal aliens, and other contraband' on the Southwest border overseen by Justice and Treasury officials but, again, no mention of the DEA (Customs, 1999).

The answer to this curious omission would seem to lie in the long history of turf battles between Customs and the DEA and its predecessors (Rachal, 1982). For example, during the reorganisation battles of 1968-73, Customs had its jurisdiction over drug cases and international presence expanded in 1972 but it did not last long. In 1973 500 Customs agents were reluctantly transferred to the newly-formed DEA; almost half resigned and many returned to Customs (Nadelmann, 1993, 140,144).

The Internal Revenue Service (IRS), headed by a Commissioner, is another Treasury law enforcement agency. In 1919 IRS established a Special Intelligence Unit to investigate fraud; this Unit was re-named the Criminal Investigation Division in 1978. Its responsibilities include the enforcement of the criminal statutes relative to tax administration and related financial crimes, violations of the Bank Secrecy Act and the Money Laundering Control Act. It says now that its main enforcement priorities include income tax evasion; fraud committed in legal industries such as health care, bankruptcy and telemarketing; and fraud committed in illegal industries such as narcotics and organised crime (www.treas.gov/irs/ci May 1999). Throughout its history the essentially *enforcement* responsibilities of the Division have not always sat easily with the broader *regulatory* and revenue-raising priorities of the Treasury. For example, Robert Kennedy sought to draw the IRS more into criminal investigations (Goldfarb, 1995, 48) but after the Watergate scandal then Commissioner Donald Alexander used the understandable controversy surrounding Nixon's use of the IRS against political opponents to effectively bring a temporary end to the Intelligence Division's targeting of organised crime and tax evasion (Block, 1991, 178-208).

The elevation of the Alcohol, Tobacco and Firearms (BATF) to separate Bureau status within the Treasury in 1972 came as there was increasing pressure on the Bureau to take a more international perspective because of demands from foreign governments to restrict the flow of weapons and explosives from the US, for example, to Mexico and Northern Ireland. However, given the small size of the Bureau and the fact that other agencies, especially the FBI, Customs and DEA, claim primary jurisdiction over the international dimensions of its mandate, its international role has been primarily supportive to those other agencies (Nadelmann, 1993, 171-74). BATF retains revenue collection functions as well as enforcement and regulatory powers, for example, under federal firearms and explosives laws it issues licenses and conducts compliance inspections. It seeks to provide expert investigative skills with respect to explosives and 'arson-for-profit' insurance frauds. Regarding firearms, its investigative priorities are said to be trafficking in weapons and their use by career criminals, drugs traffickers and gangs (www.atf.treas.gov May 1999).

Federal co-ordination

As we have already seen in chapter two, there are federal efforts to co-ordinate the transfer of information and intelligence between state and local agencies such as the RISS projects funded by Justice since 1974. Subsequent efforts have attempted to build on these, for example, in 1986 the Justice Department Bureau of Justice Assistance developed the Organized Crime Narcotics Tracking Enforcement Program that advocated multi-agency work by means of the shared management of resources and joint decision making. The formation of these 'control groups' was seen as preferable because they avoided the serious friction that often resulted from 'lead agency' management models. This idea was then incorporated in a further programme initiated by the Justice Department in 1993 aimed at developing a state-wide criminal intelligence-sharing model. After five state projects were conducted no single model was recommended largely because of the need to build on already-existing state intelligence and information systems.   Further, a survey conducted in 1996 of all states found that at least 43 states had or were planning state-wide criminal intelligence systems but that most of these would be pointer systems rather than integrated databases (BJA, 1998).

But can the federal law enforcement agencies co-ordinate themselves? Superficially, the fact that law enforcement intelligence has become more integrated within the broader structures of US foreign and security intelligence might contribute to better co-ordination.   This integration has taken place more or less parallel to the same development in the UK where, first, MI6 became involved in intelligence collection regarding drugs issues, especially in the Caribbean (Warner, 1998) and from 1996 the Security Service has had 'serious crime' added to its traditional security intelligence mandate.   In the US the same shift is symbolised by the appointment of William Webster as Director of the CIA in 1987, having served as Director of the FBI since 1978.   In April 1989 the Directorate of Intelligence within the CIA established a Counternarcotics Center which employed several hundred photo interpreters, analysts, operations officers and technicians as well as representatives on two-year secondments from other intelligence and federal law enforcement agencies (Kessler, 1992, 59). In 1994 the mission of the Center was expanded to include international organised crime and it was renamed the Crime and Narcotics Center (www.cia.gov May 1999; Raine & Ciluffo, 1994, 144).

       However, the realities of bureaucratic politics within Washington would probably indicate that while better co-ordination might be achieved with respect to *specific* multi-agency operations, *overall* co-ordination will not occur while the lion's share of operational resources and priority-setting remain under the control of the individual agencies. For example, in the area of drugs enforcement, efforts at co-ordination may stimulate rather than reduce competition.   The National Drug Intelligence Center has been established as an interagency centre with representatives from all the major federal law enforcement agencies and seeks to develop strategic analyses rather than yet another database. It has been argued that this represents one of the best of recent initiatives that should be allowed to develop a broader mandate regarding transnational organised crime generally although it is acknowledged that it was initially established as a piece of 'pork barrel' legislation and faces much bureaucratic competition (Lupsha, 1996, 44-5). Yet, the DEA, also located within the Justice Department, still maintains its own drugs intelligence co-ordinating centre in El Paso with 164 staff in 1996.   Similarly, the Drug Tsar may have achieved the co-ordination of *some* enforcement policies in *some* areas through mechanisms such as HIDTA, but it has clearly not succeeded in achieving overall co-ordination between federal agencies let alone the even more extensive fragmentation at state and local levels.

Conclusion

To European eyes, the multiplicity of public law enforcement agencies in both Canada and the US seems to suggest that the prospects for some co-ordinated 'war on crime' are remote. However, there are a number of factors that suggest that the actual difference between the European and North American situations is less significant than it might appear. For example, the actual number of forces potentially involved will not necessarily make much difference if the issue is one of information-sharing since this depends on knowledge and trust between contacts wherever they are. Potentially, sharing information between agencies *may* be as quick as within organisations - if the people concerned perceive a shared interest in some objective then they will share, if not, they will not. Actual operations will be more complex to set up the more agencies are involved, as each seeks to safeguard its own particular mandate. Certainly the multiplicity of agencies in North America place a premium on organisational politics but it does not necessarily follow that greater 'rationality' in allocating investigative resources or less 'attrition' of cases would follow from amalgamations and/or centralisation, even if they were politically likely.

ICT advances clearly facilitate information-sharing and thus the development of networks of exchange within law enforcement and, importantly, reduce the need to re-draw organisational boundaries. The problem with establishing fresh organisations as solutions to a lack of co-ordination is that those new organisations themselves may come to develop some specialised enforcement interest and, rather than co-ordinating the efforts of others, it pursues that interest either complementary to or in conflict with those others. Certainly, traditional ideas of hierarchies as providing superior forms of co-ordination become less and less relevant to the world of developing information and intelligence networks. But as the network form is developed to provide formally for information-sharing (see the discussion in chapter two) as it always has done informally, it raises profound issues for accountability and control, to which we shall return in the conclusion.

6 Targeting the Intelligence Process

Introduction

A powerful metaphor for contemporary surveillance societies is the panopticon, developed as an architecture for the prison by Bentham in the late eighteenth century and designed in order to maximise the surveillance of inmates. Subsequently developed as a more general social theory by Foucault, the modern use of the metaphor is increasingly incorporated into the discussion of the new technologies of surveillance, especially those that are electronic (e.g. Lyon, 1994). It is undoubtedly the case that the surveillance capacities of both public and private authorities have been greatly enhanced by ICT and certain aspects of these are discussed in chapter seven.

Here, the significant point to be made is this: in discussing contemporary surveillance regimes it is sometimes forgotten that, while the architecture of the panopticon provided a *potential* for the surveillance of all and thereby intended to instil self-discipline among subjects, it did not actually *achieve* the simultaneous surveillance of all. The guard in the 'inspection lodge' could only actually look one way at a time. Now, it might be argued that contemporary ICT has changed this in facilitating simultaneous surveillance of the many *via* closed circuit TV but the same principle applies. Contemporary control rooms contain monitors with the facility to show the product of sixteen different cameras on a split screen but, ignoring just how effectively one controller can observe sixteen different images simultaneously, the crucial element for policing is that the operator can select and zoom in on any one of those images.

It is crucial to distinguish two different ways in which the data gathered by surveillance can be deployed. Specific information will be gathered on individuals subject to the 'gaze' at any particular time and the significance of targeting is clear why are some targeted and not others? But, particularly with contemporary ICT, a great deal more information flows into the organisational system (or is available to it) than just that resulting from specific targeting. This material is available for analysis for a variety of purposes, including a *post hoc* search for evidence and for the search for patterns or trends as part of a risk assessment or prevention programme. Many liberal democracies in the last 25 years have shifted from 'universalistic' notions of welfare and justice towards 'new public management' and risk assessment. For example, the policies of Customs and other tax authorities have shifted towards a 'broad-based bifurcation' of their populations into 'high' and 'low' risks. For the latter, authorities will provide various forms of self-assessment and regulation subject to occasional audit, while the former (including people, firms, ports) will be targeted for more systematic investigation (e.g. Jamieson *et al*, 1998, 302-06 and see later discussion of Customs). These same trends are observable in policing though not to the same extent because a higher proportion of their workload is 'reactive'.

Thus, intelligence systems do not operate randomly. Even those that 'scan' the environment are looking for particular kinds of information. Police intelligence is more purposively directed in that its organisational mandate concerns 'crime-like events'; but given the number of such events and the small size of most CIUs the targeting process is *highly* selective. This is where the major steering mechanisms of the intelligence process are located. Priorities may be determined within intelligence units themselves, as for example, by the feedback from previous intelligence 'cycles' (↑10 in Figure 1.2, p.22) or by 'demands' from external agencies (↑1) such as other specialist police units, police managers, organised interests or ministries. The more autonomous a CIU then the more selective it may be to requests or directions from other agencies and the more likely its selectivity will be exercised in accordance with the dominant 'consciousness' of those within the Unit. This may reflect 'professional' judgements as to the relative 'seriousness' of different potential targets or stereotypical judgements as to the suitability of specific groups or individuals as potential police 'property'; though the latter will normally be presented to the public as the former.

Targeting is the start of the intelligence process and decisions taken at this stage are a major determinant of the outcome of policing. Certainly in the UK, it was the intention of those introducing intelligence-led policing that the targeting of 'known' offenders would be a more effective use of resources than reactive investigations (Audit Commission, 1993). Indeed, a trend towards investigations being 'directed' by intelligence priorities had already been noted by Dorn *et al* in their analysis of drugs policing and they predicted its further growth (1992, 148) and Sparrow's analysis of US government agencies seeking 'compliance' noted a similar increase in the identification of priorities and targets (1994, xxv-vi).

As we have already seen, law enforcement agencies investigate only a fraction of the crime that occurs, therefore agencies have extensive discretion in selecting those areas to be pursued. Police agencies vary in the degree to which they are subject to external political direction but all will enjoy some measure of operational autonomy. In a world of rational maximisation, agencies would use this to engage in environmental scanning and then target resources on those activities deemed to be most socially damaging, but the realities are that police decisions will be taken according to certain 'recipe' rules based around official discourses, organisational sub-cultures and the associated stereotyping of some groups as particularly criminogenic.

Police investigative processes are structured in order to 'get a result' in a process that has been described by McConville *et al* as 'case construction' (1991, 14-35). Similarly Ericson, in his study of detective work in Canada, emphasised the significance of targeting in 'the process by which detectives make events into crimes and people into criminals' (1981, 209). This process is central whether the reactive or proactive aspects of policework are examined. The main mechanism by which it works reactively is by 'call-screening' in which 50% or more of initial calls from the public will not be subject to any further investigation (though they will be logged and form part of aggregate databases that may be interrogated for evidence of crime patterns). The overall aim of this is to reduce the potential overload from public calls (Manning, 1992, 359). Proactively, more or less formal decisions will be taken that will result in either information- or evidence-gathering operations against specific individuals and/or groups. If this process does not encourage 'negative feedback' to modify initial targeting decisions, it may become a circular system of 'rounding up the usual suspects'.

Another way of putting this is to refer back to the twin problems of contemporary governance: knowledge and power (see discussion in chapter one). On the face of it information systems are directed to obtain knowledge and only when a decision to act is made does power become important. But if, as seems often to be the case, police believe they 'know' who has committed some crime or crimes then the objective of information gathering is to obtain not knowledge in the sense of 'finding out' something but, rather, confirmation of that belief. This is much closer to power.

Setting priorities

It must be stressed that police are faced with the structural necessity of making choices and assigning priorities even if it is only during the last fifteen years or so that they have started to publicly address the issue. Prior to that they perceived a need to uphold for symbolic reasons the myth that police could fully enforce the law (cf. Loader, 1997; Walker, 1996). Of course they could not and this failure to talk publicly about what actually *were* priorities contributed to the widespread perception of their non-accountability. The application of New Public Management to the police in UK, with its array of performance indicators, national objectives and local policing plans has changed this picture considerably but while police now proclaim what they are prioritising they remain reluctant to talk much about the areas that are consequently downgraded. In fact these 'posteriorities' are determined by a mixture of substantive ('less serious' offences) and procedural reasons (those that are screened out because the initial crime report suggests further investigation is unlikely to be fruitful).

The issue of 'viability' emerges from this study as a crucial factor in determining whether an information gathering operation or investigation will be mounted. At one level, it is entirely commonsensical that organisations do not embark on the expenditure of large resources on projects that they believe to be doomed at the outset but it raises some important questions in the context of policing. No doubt, police have always made such judgements but the issue has greater significance now. First, this is because the relatively privileged position of police within public budgets in the UK has been curtailed over the past ten years and, second, it is argued that the size and sophistication of criminal enterprises has become such that larger sums of money are necessary to fund their investigation.

In combination, these factors may give a particular twist to the paradox of enforcement that was identified in chapter one, specifically that police will be rhetorically committed to 'taking down' the most serious criminals whose operations cause the most financial or personal damage and the difficulties of doing so will be presented as reasons for enhancing the powers of police. Thus, some analysts maintain that 'One of the most important objectives in successful organized crime control is the disruption of monopoly power' since the longer the monopoly remains unchecked the more likely it can protect its position through 'alignment' with the political system (Sobocienski, 1995, 16), that is, achieve symbiosis. But when it comes to actual operations, there are other countervailing forces.

However much violence or skulduggery may have been necessary in order to establish the monopoly, once it has been achieved then that very fact may obviate the need for the use of violence other than exceptionally. This is much to the benefit of the monopolist since nothing attracts the attention of law enforcement quicker than an outbreak of violence and threats to public safety. An intelligence-led suggestion that law enforcement break up a monopoly *may* receive general support from those agencies who would be responsible for the operation but, if it was thought likely to initiate a turf war between contending groups, then there would also likely be much opposition to the plan. Further, cost and viability issues might lead them to conduct operations against less important figures where 'a result' is more likely and, consequently, the monopoly position of the major figure may actually be reinforced (cf. also Dintino & Martens, 1983, 96). This is similar to the security intelligence phenomenon identified by Brodeur in which security services, under pressure to have something to show for their efforts, concentrated their efforts on political dissent that was conspicuous and relatively easily surveilled in comparison with the more serious and covert threats that were harder to detect (1985, 7-9)

There are several areas that now tend to be prioritised as part of the movement towards intelligence-led and/or proactive policing. The first of these, especially at local level in both North America and UK is 'volume crime'. Crime Pattern Analysis (CPA) is based on the research finding that crime is seldom randomly distributed within any area (Read & Oldfield, 1995, 7). Accordingly a number of different methods of looking for patterns have emerged, including the identification of crime or incident 'hot-spots' and patterns of repeat victimisation by means of mapping occurrences and the identification of repeat offenders or a crime series by means of comparative case analysis. Although these techniques have all been recommended for use regarding local 'high volume' crime in the UK (e.g. Audit Commission, 1993) and developed most systematically by NYPD in Compstat (Silverman, 1999, 97-124), comparative case analysis has been developed furthest with regard to serious violent crime, for example, the Violent Criminals Apprehension Program in North America.

'Serious' crime, the second broad category, is actually defined statutorily in the UK. The term is used, for example, in the Security Service Act 1996 to limit the law enforcement mandate of the Security Service and is used similarly in the Police Act 1997 to limit the operations of the National Crime Squad (s48[1]) and the circumstances under which intrusive surveillance can be authorised (see further chapter seven). However, the term is defined very broadly and in such a way that it might well go beyond common-sense assumptions of what is or not 'serious':

a) it involves the use of violence, results in substantial financial gain or is conduct by a large number of persons in pursuit of a common purpose, or

b) it involves the commission of an offence for which a person over the age of 21 without previous convictions could reasonably be expected to receive a prison sentence exceeding 3 years. (Police Act, 1997, s.93[4])

Especially controversial in Parliamentary debate was the section clearly aimed at protest - note, this need not be violent - since police surveillance of those indulging in social, political or environmental protest, while convenient for authorities, may rapidly come to invade what are considered essential political freedoms. The British Government's response to such criticism was the time-honoured one of arguing that a 'rigid definition' of serious crime might hamper the authorities by restricting their operational flexibility as well as providing opportunities for unscrupulous lawyers to challenge them (e.g. see Baronness Blatch quoted in Levi, 1997, 16). As we shall see below, the NCS uses a scoring system for priorities that gives a more detailed view of how they define this term in practice.

There was a detailed discussion of the third area - organised crime - in chapter three but one of the factors discussed briefly there is a particularly controversial targeting issue. In the US issues of race and ethnicity have always been bound up with the debate about organised crime. 'Ethnic succession' was proposed as an alternative explanation for the persistence of organised crime to that of the centrally co-ordinated Mafia conspiracy that was presented to the Kefauver Committee by the Federal Bureau of Narcotics. Crime, it was argued, provided 'a queer ladder of social mobility' for successive generations of immigrants barred initially from making their way legitimately. As groups obtained a foothold in 'upperworld' they could leave the underworld: thus the Irish were succeeded by the Jews, then came the Italians and so on. The explanation has been criticised on a number of grounds, for example, that some senior crime figures did not leave crime having gained status; indeed their children often followed them into it. Also, new groups seemed to become part of organised crime, increasing its diversity, not replacing the older groups. Nor could it explain the persistence of white-collar and corporate crime that are manifestly not committed by newly-arrived immigrants (Abadinsky, 1997, 304-11; Smith, 1980, 375).

The current construction of a new European security architecture is also defined largely in terms of ethnicity and dominated by official concerns with population movements (Webber, 1993). Immigration has always been seen as a fundamental 'cause' of crime and ethnic minorities have been a major component of the 'dangerous classes' whose control has been the primary objective of police (Silver, 1967) thus it is hardly surprising that European law enforcement has adopted an ethnic classification, for example, 'Asian-', 'African-' and, most recently, 'Russian'. To the extent that new immigrant groups are less assimilated within any particular country then they represent a challenge to the ongoing process of 'state-making' (cf. Tilly, 1985, 181-84). The state will perceive the alternative national and cultural loyalties of the populations as a potential threat to their rule and their suspicion will only be enhanced by the difficulty it experiences in 'penetrating' those communities by means of surveillance.

At the most general level, 'differentiation is the relentless product of the panoptic sorting process in risk society' and the police, in a variety of ways and for a number of reasons, constantly classify people in terms of their identities, especially age, race and ethnicity (Ericson & Haggerty, 1997, 256-9). As far as practitioners are concerned, migrant and minority communities and sub-cultures provide the site for criminal activities whether organised or not and, the less the integration of the community, the more likely it is that those engaged in crime will form gangs confined within that community. However, the centrality of drugs trafficking to organised crime has encouraged inter-linking between groups based in different communities (Beare, 1996, 75-6, 82, 93-5; MacPhee, 1995; Sandelli, 1995). Still, in both the UK and North America the most regular categories in which practitioners discuss organised crime are ethnic. For example, the following comparison of the RCMP's 1995 priorities nationally and in British Columbia (see Table 6.1) differ somewhat in ordering but ethnicity is the dominant factor. The use of the term 'x-based' is intended to convey that the target is crime taking place *within* a particular community, rather than that particular ethnic groups are targeted *per se*, but the subtlety of this distinction may get lost. There is evidence that the use of ethnicity reflects police stereotyping rather than careful analysis as to the nature and causes of crime problems. In the CISC survey of organised crime, for example, ethnicity is a factor in the account only when it concerns minorities. So, in a chapter on 'Ethnic Organized Crime Groups':

Table 6.1   RCMP Priorities

Target	National	British Columbia
Hell's Angels	1	1=
Aboriginal-based organised crime	2	5
Other outlaw motor-cycle gangs	3	1=
'Traditional' organised crime	4	7
East European- (Russian-) based organised crime	5	4
Asian-based organised crime	6	3
White extremist/Hate Groups	-	6

Source: Fahlman, 1995; MacPhee, 1995

The third gang (in Alberta) is the Devil Boys. This gang has approximately 20 to 25 members and an estimated 20 to 30 associates. This group also lacks structure and hierarchy. Leadership changes on a regular basis, depending on the wealth and personality of an individual. The Devil Boys are a typical street gang of Asian descent. They loiter around restaurants, billiard halls and nightclubs maintaining a high profile and readily using violence to intimidate citizens and members of other gangs. Their primary illegal activities include break and enters, drug trafficking, prostitution, credit card frauds, assaults and extortion. (CACP, 1993, 50)

But in a later chapter on 'Youth Gangs' we are told:

Youth gang activity reigns supreme in many major urban centres cross Canada...Most youth gangs are unstructured groups with no particular leader. More dominant individuals in the gang tend to have more control over gang activity, Gang activities include assaults, sexual assaults, threats, mischief, wilful damage, drug trafficking, burglaries, robbery extortion and thefts. Youth gang members act with extreme violence using weapons such as knives, machetes, baseball bats and pipes. (CACP, 1993, 55)

So, on what basis are Asian gangs, specifically Vietnamese in the Alberta case above, singled out? Ericson and Haggerty provide further examples of racial and ethnic classification schemes used by Canadian police (1997, 282-91).

In the US it has been shown just how clearly the 'war on drugs' has targeted young blacks and Hispanics. Between 1980 and 1994 the prison population increased from 500,000 to 1.5 million (and reached 2.0 million in 2000) and the percentage of those in state prisons who had been convicted of non-violent drug offences increased from six to almost thirty. Of those imprisoned after drugs convictions 90% were black or Hispanic (Beckett, 1997, 89; see also Walker, 1999, 146-7). Occasionally the discriminatory appearance of such targeting practices becomes a matter of open controversy: the revelation that one of Metro Toronto Police's organised crime squads within Intelligence Services was dedicated to 'Black Organized Crime' (six officers, three of them black) gave rise to an instruction from the Police Services Board (the civilian oversight body) that a new protocol for Intelligence Services should be developed in consultation with the Urban Alliance on Race Relations. The resulting document sought to draw up tighter guidelines regarding such matters as the scope of information gathering, dissemination and retention of files (Metro Toronto Police, 1995) and, as we saw in chapter five, similar controversies have occurred in New York.

Some ethnic groups are apparently sufficiently well established in North America for political propriety to require that some euphemism be employed, for example 'Traditional Organized Crime' is now used instead of what used to be called 'Italian': the Sicilian Mafia, La Cosa Nostra and the 'Ndrangheta linked with Calabria (e.g., CACP, 1993, 43-5). Similarly, it was suggested that 'Hate Crime' became a priority for the Metro Toronto Intelligence Services not because the police identified it as such in terms of 'seriousness' but because of political pressure (Sandelli, 1995). In the UK, the report of the Macpherson (1999) Inquiry identified 'institutionalised racism' as a major factor in the failed Metropolitan Police investigation into the murder of Stephen Lawrence. The identification of how racist stereotyping permeates a variety of police investigative practices is clearly of significance when considering targeting for covert information gathering.

Targeting in Canada, UK and US

Within both the provincial and federal structures in Canada, priorities are established in an apparent 'bottom-up, top-down' cycle that ultimately permits a great deal of local latitude. For example, full members of CISO are asked to identify issues of concern in their communities, from which a list of anything from four to a dozen provincial problems are identified. These are then transmitted to CISC which draws up a list of national priorities - the main criteria being, unsurprisingly, the extent to which a problem is countrywide rather than local. When the national priorities are sent back to the provinces each will concentrate upon those of most pressing provincial concern (Faul, 1995; Levesque, 1995).

The dominance of local concerns raises questions as to the utility of developing national priorities at all. At the national level this seems to be primarily presentational, for example, each year the major public product of CISC was the report it wrote for publication by the Canadian Association of Chiefs of Police (CACP) Organized Crime Committee until it was disbanded as superfluous to CISC in the Spring of 1995 (Beare, 1996, 31-2). This report sought to provide a survey of the extent of organised crime without giving any indication of what the priorities were. From the point of view of provincial and local forces the main utility of the national priorities is that obtaining funding for operations is facilitated if they are consistent with some national priority (Faul, 1995; Sandelli, 1995).

Within the RCMP a similar process takes place: each operational area and project in the divisions report on their activities every six months and identify immediate and longer-term issues. The Criminal Intelligence Directorate add in their own strategic review and proposals are then presented to the Criminal Intelligence Steering Committee who, in establishing national priorities, ask themselves the following questions:

- what benefits are to be derived or what practical significance to the RCMP does the project have?
- what is the level of threat/harm to society?
- are there political/legal implications?
- what is the cost versus the benefit?
- will the focus of intelligence go beyond low-level criminal operations?
- are the financial, materiel and investigative/analytical resources and access to information in place to undertake the project?

- will the potential results have long range strategic benefit in terms of future planning, policy development and resource allocation? (RCMP, 1991b, 8-9)

It is not made clear whether these are in order of priority; if they are, putting benefits to the RCMP above the level of threat to society is an excellent example of the 'goal-displacement' that characterises bureaucracies!

As would be expected, the size of CIUs affects their clout: smaller agencies who are less able to contribute to joint operations will feel that they benefit less (Walsh, 1982, 13) while the larger ones, for example, Metro Toronto within CISO and the RCMP's Division in British Columbia are better able to influence the priority-setting process in the first place and then to actually pursue local/provincial objectives autonomously if they so wish (MacPhee, 1995; Sandelli, 1995). Smaller forces in the UK similarly felt a lack of influence compared with larger urban forces when it came to Regional Crime Squad (RCS) operations - this feeling is not likely to have been ameliorated by the merging of the RCS into the National Crime Squad.

Interviews with practitioners revealed a fairly uniform set of general criteria for prioritising specific criminal groups or activities but in only a few cases was a more systematic methodology employed. For example, the RCMP was seeking to develop such a system (Fahlman, 1995) and the NCS used one (see below). While some thought this would be useful, others thought it might just inhibit the normal processes of intra- and inter-agency exchange that preceded targeting decisions. Certainly, in some cases, the room for manoeuvre permitted by stated objectives was so broad as to render them almost meaningless. For example, twenty years ago a CISO Workshop on Organized Crime concluded, *inter alia*, that their definition of organized crime as:

> two or more persons consorting together on a continuing basis to participate in illegal activities either directly or indirectly for gain is *TOTALLY* inadequate and requires immediate revision. (CISO, 1979, emphasis in original.)

Yet, in 1993 both the RCMP (1991a, 23) and CISC (Levesque, 1995) still used precisely the same definition. Such a broad definition has the advantage of maximising the discretion of practitioners in deciding where to deploy resources but it lays the way open for the incorporation of almost any criminal enterprise, however unstructured or trivial.

The overall priorities of law enforcement agencies tend to be established out of a process of negotiating various influences, for example, legal and geographical jurisdiction, political guidance and relations with other agencies. The Manhattan District Attorney refers to Morgenthau's general criterion of 'economic and social viability of the City' (Castleman, 1999). Within the parameters of these priorities, however, no New York agency claimed there was any specific methodology for establishing targets and specific investigations are primarily reactive, for example, to a request from another agency, a good piece of information from an informant or as 'spin-offs' from previous investigations.

When it comes to assessing which of any number of groups will actually be targeted there is general agreement on the relative seriousness of different factors. 'Doability' and the likelihood of success is always a factor and the more specific the allegation the more likely it is to be investigated. Also, having an inside source was seen as dramatically increasing the prospects for success. Similarly, cases are more likely to be pursued if they involve group rather than individual activity and, within DOI, if they are part of pattern within an agency (Burke, 1999). Most interviewees identified violence (especially in connection with controlling territory) as either the top priority or as a factor increasing the priority of any target. The priorities of the Homicide Investigation Bureau in the Manhattan DA's office are set on the criteria of violence - information gathered from informants and elsewhere is evaluated with a view to identifying 'the worst of the worst' gangs (Arsenault, 1999).

The next most frequently mentioned factor in New York was the corruption or subversion of the political process, for example, *via* influence-peddling; then issues of financial impact, including money-laundering. The targeting of allegations of corruption is a sensitive issue in all agencies: the Manhattan DA and the DOI spend most time on the low-level corruption amongst the thousands of inspectors and other officials rather than with politicians. This is partly because cases are easier to win: pay-offs regarding permits, contracts and inspections are easier to prove than the legally more ambiguous area of politicians 'doing favours' for constituents (Castleman, 1999). It could also be said that this will bring less political heat on the agencies.

In the UK the South East Regional Crime Squad developed a scoring system for operations. This was an attempt to systematise the process by which the Squad would decide its priorities among the various 'jobs' submitted to it by its constituent forces. This was part of the negotiation between three previously separate RCS when they were merged and it was hoped it would lead to a better understanding of Squad priorities and a better allocation of RCS resources (Bliss & Lowton, 1995). Subsequently the system was adopted by other RCS and, when these were merged in 1998, by the National Crime Squad.

Initially under this system for any operation to be accepted by the Squad it had to achieve an overall score of eighty points (see Table 6.2). In calculating scores, no more than fifty points could be carried forward from any one column, for example, 'quality of criminal' but it can be seen that eighty points did not reflect a very high threshold given that anywhere between twenty and fifty points could be scored on each of seven variables. The threshold was increased in the NCS Service Plan for 1998/99 that identified one of its Performance Indicators as achieving a minimum score of 100 points in 80% of the operations it accepts (NCS, undated, 9). Since, during the second half of 1996 all the targets selected had averaged over 120 points (Penrose, 1996), even this higher threshold was hardly demanding! It is not known how rigidly or otherwise the system is used in determining priorities among those potential operations that do pass the threshold but it provides interesting insight into the judgements made by senior police as to relative crime seriousness.

One stated aim of the system was that it should not be restricted just to drugs operations but it can be seen how the system pushes police towards drugs operations; any significant drugs importation could quickly acquire a high score and, in the initial evaluation of the scheme, 60% of new cases were found to be drug-related (Bliss & Lowton, 1995). Otherwise the system reflects the predominant concerns of the national level agencies with elements of organised criminality and international dimensions. The 'core' and 'current' nominals referred to under 'Quality of criminal' relate to the different NCIS method of prioritising its resources. The Service initially used a more general set of criteria in the construction of its nominal index but in its first few years this grew so rapidly that it feared becoming just a 'library' of information and so during 1996 it developed these criteria into a scoring system that led it to identify two groups of targets: 'core' and 'current' nominals of whom there were respectively 151 and 2,358 in 1998 (NCIS, 1998a, 8).   These systems relate to *tactical* or investigation priorities. Currently, NCIS is also seeking to develop a more systematic method of determining *strategic* priorities.   At its simplest, this seeks to employ environmental scanning for threats that are then scaled in terms of high to low with reference to three criteria: financial (amount of money

Table 6.2   **National Operations Evaluation Formula**

(adapted from Bliss & Lowton, 1995)

Points	Quality of Criminal	Quality of Intelligence (4x4)	Victim
50	Heads criminal organisation or core nominal	-	Serious threat to life
40	-	A1	Serious potential of significant personal harm
30	Significant member of criminal organisation or current nominal	B2	Serious impact on UK economy/ European banking system
20	Peripheral member but arrest will disrupt either of above	C2	Serious loss (or potential) likely to jeopardise company or individual livelihood aggravated by multiple victims/ offences
15	-	C3 or B4	-
10	-	X3	Serious loss (or potential) likely to jeopardise company or individual livelihood
0	-	X4	-

Table 6.2 (continued)

Profit from Criminal Activity (potential in 6 month period)	Type of Offence	Geographical Impact of Operation	Quantitative Measures for Drugs
-	Kidnap; Incitement/ conspiracy to murder	-	-
-	Product contamination	-	-
-	Serious Firearms offences	International	-
£1 million	Major drugs trafficking; money laundering	National	>5kg heroin/cocaine >10k doses of LSD/ ecstasy >1 tonne cannabis >20kg amphetamines
£0.5 million	Other series or serious crime requiring RCS capability for best resolution	Interforce	>1kg heroin/cocaine >5k doses of LSD/ ecstasy >0.5 tonne cannabis >10kg amphetamines >100gms crack
£0.25 million	High value lorry load; Forgery/ counterfeiting	Within force requiring RCS capabilities	>0.5kg heroin/cocaine >2.5k doses of LSD/ ecstasy >100kg cannabis >5kg amphetamines
-	-	-	

being diverted from legitimate, intended purposes and used to profit the criminal), societal (the scale of social concern or degree of protection sought by society at large over a period of time) and political (the degree of urgency or imperative for action locally or nationally). The resulting matrix might look like that in Table 6.3.

Given that such a matrix could be increased in sophistication to allow for the inclusion of more variables and subtle distinctions, then it may well provide a useful basis for the strategic assessment of threats but it is unlikely that it could achieve the desired aim of avoiding the need for law enforcement agencies to negotiate priorities around a table (Hall, 1999, 14). It *might* be possible to achieve some consensus around a scoring system but it could never replace negotiations between agencies because their differing structural relations to government, mandates, areas of operation and resources will ensure that they may well legitimately interpret the 'scale of social concern' or 'degree of urgency' quite differently.

Table 6.3   Strategic Threat Assessment

Criteria/scale of threat	High	Medium	Low
Financial	£,000 millions	£,00 millions	£,0millions
Societal	national	regional	local
Political	international	national	regional/local

Source: Hall, 1999, 14.

There are various mechanisms in UK Customs for determining priorities, with the National Intelligence Division (NID) itself involved in providing information to the committee that determines the allocation of Customs resources between the main business areas of prohibited goods and revenue (Wesley, 1997). A model is being developed for the allocation of the resources of the intelligence division itself that is based on the idea of risk assessments. Relating to drugs, for example, the formula builds in factors such as the previous incidence of seizures and class of seizures per numbers of passengers at the different ports of entry. Within the broad remit of national priorities, regional Intelligence Co-ordination Units (ICUs) appear to enjoy considerable autonomy. Requests from operational teams and investigators provides 90% of the work of intelligence staff and the ICUs encourage this 'customer-led' approach to their work. When ICUs are approached they will seek to develop a package that enumerates the inputs and outputs of the proposed operation in an attempt to put a cost-benefit value on it before a decision is taken. This approach is similar in intent to the NCS points system but, as ever, clearly leaves room for the important informal contacts and negotiation between officers. Assigning priorities means, of course, that many potential jobs remain undeveloped - in Customs one estimate of the number in this category was 70%.

Within North Wales Police procedures for identifying priorities for intelligence-development were still evolving, at the time of interviews early in 1997. Formally, force priorities were determined by a Tasking and Co-ordination Group (TCG). But this did not always meet at the specified fortnightly intervals and between times proposals would go from the Force Intelligence Bureau to the Detective Chief Superintendent for decision as to whether particular operations would be taken forward and, if so, by which division or specialist squad. There was a general sense that the formal process was slow and could be short-circuited if necessary.

Practice varied somewhat at divisional level, in one, 'crime strategy' meetings would discuss each month 7-8 packages that had been developed by the Local and Field Intelligence Officers and decide what could be dealt with within the Division or whether it should be sent to the FIB because it needed greater resources. In another division fortnightly 'tasking and co-ordination conferences' were held. The degree of autonomy enjoyed by different squads to determine their own priorities also varied, for example, the force drugs squad apparently enjoyed considerable leeway given that most of its work was initiated by squad officers themselves. On the other hand the force's Serious Crime Squad operated entirely reactively; for 3-4 years prior to 1997 the entire squad had been involved in inquiries in relation to a public inquiry into child abuse in North Wales children's homes and the only crimes that had taken the squad away from that had been occasional murders.

Within the main priority areas - local 'volume' crime within divisions, cross-border or 'professional' crime at force level - the first criteria used by senior officers to identify specific targets was some notion of viability and any information gathering initiated would be reviewed at intervals in term of continuing costs and what was being achieved. Other things being equal, the perceived quality of information received and the presence of violence would increase the priority of any target. Given the child abuse inquiry and the murder of a child - Sophie Hook - in Llandudno in 1995, it is not surprising, perhaps, that particular attention was focused on paedophiles.

Barton and Evans found a similar variety of practice in holding tasking meetings across different areas in Merseyside and argue that the holding of regular, formal meetings attended by representatives of all ranks and teams was a significant factor in the successful integration of all officers into the model of intelligence-led proactive policing being introduced there (1999, 30-31). Silverman notes the importance of regular meetings to the Compstat process (1999, 102-03).

It was striking how frequently police interviewed referred to the fact that their first enquiry on hearing of a specific crime or crimes would be to seek information on recent prison releases of people who had been convicted previously in their area. This seems to be one of the dominant 'working rules' of criminal investigation. If that enquiry proved fruitless, then a more systematic search of relevant name *and modus operandi* indices would be required. It is precisely this kind of targeting of 'known' criminals that the Audit Commission (1993) sought to encourage. But if policing is driven *solely* by the targeting of 'known' criminals there is clearly a danger that the whole process becomes little more than the endless re-cycling of a specific group of people around the criminal justice process. Therefore, the question was asked as to what efforts were made to identify previously unknown offenders. The answer was - very little, mainly because there were enough 'knowns' (believed by police to be disproportionately responsible for local crime) to keep everybody busy. The only way in which a hitherto unknown offender might be produced is by way of crime pattern analysis that suggested specific locations for surveillance and investigation.

Conclusion

It is clear from this study that the targeting priorities of law enforcement agencies vary largely in relation to their structural location - 'what you see depends on where you sit'. The process of selecting them is pre-eminently a matter of organisational politics, notwithstanding attempts to inject more systematic schema into the discussion. In national-level agencies such as

NCIS and the DEA the focus is primarily on transnational criminal operations. Regional agencies, such as the former RCS in the UK and state/provincial agencies in the US and Canada are concerned with domestic crime that crosses internal force borders and local agencies are concerned with operations primarily within that area. This is barely surprising but its significance is not always taken into account in discussions of police co-operation and co-ordination, whether by practitioners or academics. Too often, failures of co-operation are presented as the consequence of ignorant parochialism (on the part of local agencies) or irrelevant supra-nationalism (on the part of national agencies). To be sure, simple bureaucratic jealously may be at the root of some of these problems but there are more fundamental structural issues at stake, including the quite proper concern at the security of information passed on. Perceptions of what 'local' enforcement priorities should be can differ perfectly legitimately and need to be negotiated. The idea that these problems can be overcome by the creation of some over-arching police force or mechanism for the establishment of priorities is a fantasy.

Some efforts are being made to systematise the targeting process. If this increases the transparency of a process which is so crucial to the impact of law enforcement intelligence, then it is to be welcomed. But it is not clear that this is happening; the systems in place operate in a way that leaves much room for manoeuvre in negotiating targets within and between agencies. Therefore, to the extent that targets are chosen more in line with police 'recipe' knowledge of 'criminals' in their area, then targeting will determine what intelligence is gathered and will not necessarily itself be an 'intelligence-led' process. For that to happen, the targeting process must be open to 'negative' feedback on previous policies and practices.

7 Technologies of Information Gathering

Introduction

There are three primary types of information that can be gathered in order to develop intelligence. First, there is the store of information within the agency or elsewhere in the law enforcement community. Given efficient means of handling the internal information flow, the former is immediately accessible, gaining access to the latter may be more difficult. Second, there are a variety of more or less 'open' sources including the records of other state and private sector agencies for investigative purposes and an infinite variety of wide-ranging social and economic data if the objective is strategic intelligence. Third, there is the information that can be collected only by covert means, either technical (for example, interception of communications) or human (informants or physical surveillance). Human sources are discussed in chapter eight; the others are considered in this chapter.

The legal context

There is not the space to provide a detailed examination of the different legal regimes regarding information gathering in Canada, UK and US but it will be useful to have some general understanding of the main similarities and differences, especially regarding covert surveillance. These regimes are a function of the extent to which general privacy rights have been built into the rules governing police investigation. These are most extensively developed in the US as a result of the fourth amendment to the Constitution that prohibits unreasonable search and seizure and was applied increasingly to police from the 1960s. At that time the situation in both Canada and the UK was still governed by the common law in which the narrow association of privacy rights with property meant that there were rules requiring police to obtain judicial warrants in order to search property but not for aural or visual surveillance (whether or not it required access to property). Change came first in Canada when the Protection of Privacy Act was passed in 1974 and accelerated once the Charter of Rights and Freedoms was enacted in 1982. In the UK change came more slowly and erratically, first, in response to certain state practices being outlawed under the European Convention on Human Rights (ECHR) and, later, as authorities sought to amend the rules in anticipation of future challenges under the ECHR or Human Rights Act.

Information gathering may be subject just to internal controls or also to external authorisations or audits. Internal controls tend to follow the principle that the greater the intrusion the higher the authorisation from within the agency, for example the NCIS codes published in the UK in September 1999 establish such 'ladders' of authorisation. External regimes are of four main types: prior authorisation by a judge, minister or official; the enforcement of guidelines by the judicial application of an exclusionary rule; pre-trial decisions determining defence access to the records of how an investigation was conducted (cf. Field & Jörg, 1998, 324 - the issue of disclosure is considered in chapter nine); and, fourth, *post hoc* review by commissioners and or tribunals as in the UK.

In 1928 the US Supreme Court held 5:4 in the *Olmstead* case that wiretapping was not a violation of the Fourth Amendment because it involved no physical intrusion on the defendant's property nor was anything seized (Abraham, 1982, 140-6 on which following based.) Although the 1934 Federal Communications Act (and some state statutes) expressly forbade it, the FBI initially continued wiretapping on the grounds that the Act did not apply to federal investigators and that interception alone was not a violation (it became so only if it were accompanied by divulgence). Further Court rulings in 1939 under-mined this position and there was a brief ban on FBI wiretaps until 1940 when Roosevelt's secret wiretapping directive to the FBI in cases of 'subversive activities' effectively opened the floodgates since Hoover was required to account to no-one for his authorisations (Theoharis & Cox, 1989, 169-71).

By the 1960s advances in technology had brought the use of electronic devices into widespread use. State law in New York authorising their use by law enforcement if approved by a court was declared unconstitutional by the State Supreme Court in 1965 and this was ruling was substantially upheld by the US Supreme Court in *Berger v. New York* 1967. In the same year, however, the Court sought to facilitate the use of the technique while adding safeguards. In *Katz v.US* the majority found electronic surveillance to be constitutional if carried out with a court warrant and that this applied also to its use in semi-public places such as phone booths. Congress followed this in the context of the rise of 'law and order' as a political issue by enacting Title III of the Omnibus Crime Control and Safe Streets Act (OCCSSA) in 1968 that permitted court-approved inter-ceptions by both federal and state law enforcement officials of a large number of crimes. An exemption was created so that, in emergencies relating to 'national security' or 'organised crime', interception was permitted for 48 hours while application was made to the court. The Government's claim that the President's inherent powers meant that national security interceptions against domestic targets could be conducted without a court warrant was rejected unanimously by the Supreme Court in 1972.

Subsequent developments in the US were mainly the result of the revelations through Court actions and congressional inquiries as to the extent to which federal , state and local law enforcement agencies had used intrusive surveillance techniques, primarily in the gathering of political intelligence, without court warrants. This led to a dramatic curtailment of FBI 'security' investigations in the mid-1970s and the enunciation of new Attorney General's guidelines setting out the conditions and procedures for such investigations (Elliff, 1979). At state level there were a variety of court-led initiatives aimed at restricting police to intrusive surveillance only after they had established that there was 'reasonable suspicion based upon specific and articulable facts that the individual or group is involved in criminal conduct...' (Lardiere, 1983, 1036; also Donner, 1990, 345-79). In New York now wiretaps can be conducted only with the authority of a court and warrants are granted for thirty days at a time, the judge receiving weekly up-dates on progress. Renewals can be applied for. Once a tap is finished then the subject must be told of it thirty days later; however, there is provision for police to go to court to get a stay of this requirement.

One interesting, and possibly unintended consequence of these restrictions on the use of technical means of surveillance was to contribute, along with various other legal, organisational and technological factors, to the increased use of police undercover operations in the US. These other factors included the exclusionary rule, the privilege against self-incrimination and restrictions on inter-rogation, privacy legislation that inhibited other agencies from passing on information to the police and freedom of information legislation that, the FBI claimed, made it much harder to recruit informants (Marx, 1988, 46-54; undercover policing is discussed in chapter eight).

In Canada a 1969 Report of the Canadian Committee on Corrections chaired by Roger Ouimet discussed the issue of technical surveillance. Its report proposed that prior judicial authorisation be required and this was implemented in the 1974 Privacy Act. As with OCCSSA in the US, the Act enumerated a catalogue of offences viewed as sufficiently serious in themselves or particularly appropriate for technical surveillance plus the residual category of 'organised crime'. After the adoption of the Charter of Rights and Freedoms the Canadian Supreme Court adopted an approach to these matters premised on the 1967 *Katz* decision in the US, that is that the Charter protects an individual's reasonable expectation of privacy and requires a balance between that and the government's interest in law enforcement. The Canadian Law Reform Commission's later review of the law in this area found that electronic eavesdropping was being used much more than had been intended, and significantly more than in the US. Their Report noted judges' reluctance to apply conditions to authorisations and that almost all included 'basket' clauses (failing to identify specific targeted names or premises). Although the legislation adopted the exclusion of illegally gathered evidence as the fundamental sanction on the authorities, the common law background of the judges had led them to be reluctant to use this (Law Reform Commission, 1986, 1-12).

Following in the empirical mould of English law, the law relating to police use of covert information gathering techniques has developed piecemeal in the UK and now provides an unsatisfactory mixture of measures that have, in recent years, developed quite rapidly. Based on the lack of any general right to privacy and that whatever is not prohibited by law is legal, the main historical restriction on the police's ability to gather information covertly has been the resource constraints. If there was any legal challenge to state practices then the prerogative power was assumed to provide the necessary authorisation. But in the last twenty years there has developed a pattern in which governments have been obliged either by losing cases in the European Court of Human Rights or anticipating future losses to introduce new legal frameworks governing intrusive surveillance.

This happened first regarding telephone tapping when, in the case of *Malone*, the ECHR found that the UK practice of authorisation by ministerial warrant with no procedure for public challenge was contrary to the European Convention of Human Rights (e.g. Fitzgerald & Leopold, 1987, 133-54). In order to legalise UK practice therefore the Interception of Communications Act (IOCA) was passed in 1985. This outlawed interception (both telephone tapping and mail) unless done with the authority of a ministerial warrant. IOCA also established a complex procedure by which a (judicial) commissioner could check that warrants were properly authorised and a tribunal (of three lawyers) could investigate public complaints. This was a procedure designed to achieve minimal compliance with the *Malone* decision as evidenced by the fact that the Commissioner's review was essentially procedural (based on the *Wednesbury* principles) and it was not part of the job to investigate whether unlawful interceptions had been carried out (Lloyd, 1990).

Further anomalies arose when the Security Service Act 1989 and the Intelligence Services Act 1994 authorised 'interference with property' by the Security Service, SIS and GCHQ. Without saying so explicitly, these permitted the kinds of property trespass often associated with electronic surveillance. The lack of an equivalent legal framework for police was criticised by the House of Lords in *R v Khan* 1996 which, together with the fear of further European challenges, motivated the government to provide such a framework in the Police Act 1997. Again, however, no attempt was made to provide a comprehensive regime for regulation of police surveillance and the use of the product as evidence; rather it was limited to protecting the police from civil actions or further European challenges (Uglow & Telford, 1997, 59-60) but it was not clear that it would withstand challenge under the Human Rights Act (Uglow, 1999). This seems to be precisely what the Government had in mind when it introduced in 2000 its Regulation of Investigatory Powers Bill to provide a more comprehensive framework including modifying the Police Act (see further below).

But before the publication of this Bill, in September 1999 the law enforcement community had sought to head off such challenges by publishing five codes of practice (NCIS, 1999a-e) together with a declaration that set out the case for covert techniques, the ethics and standards to which it subscribes and rights of redress (NCIS, 1999f). The codes seek to plug the gaps left by the piecemeal nature of the legislation to date, specifically by providing for a 'ladder' of internal authorisation of covert techniques in those cases where either ministerial (interception of communications) or judicial ('interference with property') authorisations do not apply. The specific codes are examined below in connection with the discussion of the technique to which they apply; here some of the general principles enunciated in the declaration are considered.

Although some of the techniques described in these codes are directed at 'the prevention and detection of crime' and anti-social behaviour in general, the major thrust of covert techniques is said to be aimed at 'serious crime and organised criminality'. In the declaration these are described as 'corrosive of civilised society' (NCIS, 1999f) and their growth is said to be evidenced by the fact that Parliament and the courts have provided law enforcement with increased powers to use covert techniques and to protect them from exposure by the application of public interest immunity. It is worth noting that the law enforcement agencies themselves are the primary providers of information to ministers, Parliament and the courts on these matters so there is an element of circular reasoning here.

The declaration is, at least, refreshingly free of the rhetoric of 'wars' against crime and/or drugs regularly used by ministers. Rather, it is said, the agencies seek to tackle the *consequences* of serious and organised crime by way of 'mitigating' the damage of criminal behaviour to the environment, the quality of life, the administrative and economic infrastructure and public confidence in the criminal justice system and the rule of law. The basis for achieving this is 'the ability to describe and analyse the nature of the criminal problem', specifically, criminal motivations, the extent of networking, the organisation and interdependence of criminal people and groups, the methods of criminal businesses and their market positions.

The 'ethical precepts' outlined in the declaration are brief: individuals will be targeted only when it is justified, authorised and the information gathered will be properly recorded and only retained or made accessible where legitimate need can be established. The declaration sets out the standards for justification, proportionality, necessity and accuracy. For the first of these the agencies

> will establish "sufficient cause", based on a suspect's previous criminal history or on reasonable suspicion of criminal activity or association, before collecting and recording personal information on the suspect in intelligence systems. (NCIS, 1999f)

This shows quite clearly how the 'net' for information gathering is set even more widely than for the exercise of police powers of stop and search or arrest where, formally, 'reasonable suspicion' is required that someone is carrying stolen or prohibited articles (Police and Criminal Evidence Act s.1[3]) or has committed, is committing or is about to commit an arrestable offence (PACE s.24 [6-7]). Research has indicated how, in practice, reasonable suspicion provides little protection against the deployment of these powers in situations where it is 'useful' for police rather than being strictly justifiable in law (e.g. Bottomley *et al*, 1991, 12, 47), in part because of the lack of a check over their abuse, but no suspicion of a specific offence is required for information gathering. Indeed, there may not be any suspicion of *current* criminal activity at all - a suspect's previous criminal history may provide the trigger. This is certainly a factor looming large in the 'working rules' of police - as we saw in chapter six the first reaction of police to, say, a series of burglaries in an area, is to ask whether any locals with a record for burglary have been recently released from prison.

A statement that covert techniques will be applied only where criminal activity is sufficiently serious to justify the degree of intrusion into privacy that it would entail is to be welcomed but there is no further indication of how this will be implemented. The ability to make judgements seeking to balance privacy rights with those of law enforcement has not hitherto been considered an especially important qualification for middle-ranking police officers if only because English law has contained no generalised right to privacy. How extensive will be the effort to train these officers within the new context of the Human Rights Act? But even after any such education programme, within police forces endowed with a pragmatic, results-driven culture and subject to cash limits and performance indicators, issues of 'proportionality' are likely to be determined more by matters of cost than by debates on ethics.

Under the principle of necessity, covert techniques should be applied only if other means are unlikely to be successful. This does not mean that all other means must have been tried and failed but, rather, that in the circumstances, they would be unlikely to be successful (e.g. NCIS, 1999d, note 1b). This principle existed in the previous Home Office guideline to police regarding electronic and visual surveillance (1984) but the wording is interesting. It does not talk about *information* that cannot be gathered by other means as necessitating covert techniques; rather this applies 'where it appears that what the *action* seeks to achieve could not reasonably be achieved by other means' (emphasis added). At one level, this may be taken simply to indicate that police and Customs are not simply 'intelligence' agencies in the sense that that they gather information, develop intelligence and disseminate it. They are executive agencies who may also act by means of arrests. But 'action' may also include a variety of means other than arrest, such as measures taken to disrupt criminal activity or organisations (see further below in chapter nine).

Notwithstanding the police's statement that they did not believe legislation to be necessary, the Government has proposed just that. In February 2000 it introduced the Regulation of Investigatory Powers Bill that, for the first time, seeks to encompass police, intelligence and armed services. The first part seeks to up-date the Interception of Communications

Act 1985 to take into account such developments as mobile phones and the Internet. It retains essentially the same procedures, that is, the requirement for ministerial warrants authorising interception. Part III seeks to deal with the problem of encryption on the Internet by the issue of notices to service providers forcing the disclosure of codes. The sanction for refusal to comply with a notice is up to two years' jail; if the notice includes a secrecy provision and the person served makes an unauthorised disclosure of the fact of the notice to another person then there would be an additional offence carrying up to five years' imprisonment. Whether this attempt to maintain a traditional form of sovereign state control will be successful over the more libertarian 'nodes' on the Internet remains to be seen.

Part II covers surveillance and human sources. It distinguishes between 'directed surveillance', for example, targeting of people that is not intrusive; 'intrusive surveillance', covering residences and private vehicles; and 'covert human intelligence sources'. As far as police and customs are concerned, the second of these is already covered by Part III of the Police Act 1997 and intelligence services have been required to obtain warrants for 'interference with property' since 1989 (MI5) and 1994 (MI6, GCHQ). This is the first time, however, that the vexed issue of informants has been dealt with in a statute. Their use may be authorised if necessary and proportionate in relation to a long list of circumstances including national security, crime, disorder, public safety, tax collection and any other purpose specified by an order from a minister. Also there are requirements in terms of handlers, supervisors and secure record-keeping (see detailed discussion of informants in chapter eight).

The accountability structures established in Part IV of the Bill are essentially extensions of the existing mix of commissioners and tribunals. An Interception of Communications commissioner will take over from the former IOCA Commissioner and a new Covert Investigations Commissioner will be set up to review the new surveillance powers, except where these are already covered by the other commissioners already established by the Security Service Act, Intelligence Services Act and Police Act! A Tribunal is to be appointed, as usual this is to consist entirely of lawyers and will investigate complaints. However, the new dimension to the Tribunal's work is that it will act as the Tribunal under the Human Rights Act to hear challenges that state actions were incompatible with Convention rights. This is potentially significant, though the Government have sought to confine the Tribunal's remit to the principles of judicial review, that is, making essentially procedural judgements as to the 'reasonableness' of the original authorisation for surveillance in relation to necessity and proportionality.

The 'knowledge store'

The first question to be asked regarding police information gathering is what use police make of their organisational memory. Simple considerations of efficiency and economy suggest that agencies should first consult their own 'store' of knowledge before embarking on further investigations. The organisational memory is what enables the organisation to avoid simply reacting to current information and demands from outside; in more general cybernetic terms it is the memory store that is the basis of the system's autonomy and which enables it to re-direct its own goals (Deutsch, 1966, 128-9, 206-7).

Where central indices and files are computerised then the ability of agencies to 'mine' them is enhanced.   In Canada, the municipal and provincial members of CISC were until 1995 dealing with three main data bases: their own, the provincial one, and the national Automated Criminal Intelligence Information System (ACIIS). The last of these was to be replaced in 1996 by windows-based ACIIS II which is intended to integrate these three levels, for example, each province will have a separate 'client server' to facilitate direct access for more people than hitherto. Not wanting to wait for the development of ACIIS II, in 1992 CLEU purchased its own system with, as of April 1995, 19,000 subjects on file (CLEU, 1995b, 2). The advantage for CLEU was that this was a superior system for analytical purposes but, as a stand-alone system, other agencies did not have automatic access (Organized Crime Independent Review Committee, 1998, 28).

The RCMP's Criminal Intelligence Program makes use of two main data bases: the National Criminal Information System and the Secure Criminal Information System (SCIS). The first includes the usual indexing and free-text searching facilities plus a capacity for maintaining separate encrypted terminals for information requiring a higher degree of protection. SCIS was designed for the storage and analysis of national security information but the spare capacity that became available in the mid-1980s when the RCMP Security Service was civilianised and became the Canadian Security Intelligence Service (CSIS) has now been made available for any investigation requiring specially secure conditions (RCMP, 1995, 7-9).

The continuing separate development of both the CISC and RCMP systems raises issues of efficiency and 'value-for-money' that cannot be answered by this brief survey. As of 1995 there was certainly evidence of organisational politics, for example, concern in the RCMP at the lack of encryption of information within ACIIS and surprise at CISC that the RCMP had not then requested access to ACIIS2 (Fahlman, 1995; Levesque, 1995). On the other hand, in the small world of law enforcement intelligence, the extent of informal contacts between people and the existence of secondments and Joint Force Operations all contributed to a picture of much easier information-sharing than an examination of the formal agreements between agencies would suggest. Of course, one of the effects of this is to minimise the inhibition of the Privacy Act 1983. Although this exempts criminal intelligence data, should such data end up becoming public or finding its way into an affidavit or evidence then formal sharing of information between agencies might become subject to scrutiny. But if, for example, members of different agencies are working on a joint investigation, their informal interaction and sharing of information is effectively beyond the reach of the legislation (Faul, 1995).

The problem of duplication came to a head in British Columbia. As we have seen, in June 1998 the Provincial Attorney general set up a review of the effectiveness of the agencies in combating organised crime and this concluded with the establishment in March 1999 of an Organized Crime Agency with an RCMP Assistant Commissioner as its chief officer. The review centred on the alleged shortcomings of CLEU, for example, that the co-ordinated response it has originally provided, had been lost and that its uneven leadership, lack of accountability and workplace conflict had undermined morale (Organized Crime Independent Review Committee, 1998, vi). These problems had been specifically reflected in the development and use of the data bases described above. CISBC decided to enter its intelligence into CLEU's system rather than ACIIS which meant that organised crime intelligence in British Columbia was not available to forces elsewhere. Further problems arose from the fact that, even within CLEU, its own Outlaw Motorcycle Gang Unit, citing security concerns, maintained another stand-alone system (Organized Crime Independent Review Committee, 1998, 28, 41-2). With the demise of CLEU, the Review recommended that CISBC be the lead agency for organising gathering and analysis, that ACIIS II be the centralised computer store and that all law enforcement agencies be directed to record information there or, at least, provide pointers to other data bases holding information (Organized Crime Independent Review Committee, 1998, 63).

In New York the difficulty of obtaining suitable software programmes and inadequate resources meant that a number of agencies, including NYPD, were not really benefiting from new technologies. Prosecutor's offices have also been slow in developing ICT: neither the Manhattan DA nor Special Narcotics Prosecutor had much by way of computerisation - one interviewee referred to the store of memory as being 'old-fashioned' though claiming that part of the reason for this was security regarding on-going investigations and the desire to encourage 'need-to-know' in offices to prevent leaks. The Manhattan DA is seeking to develop a case-tracking system that will help avoid one investigation from 'contaminating' another by, for example, granting immunity to a witness who is a target in another investigation (Castleman, 1999). The information store within the Department of Environmental Conservation police is primarily paper records; there is just a simple database system based on Microsoft Access that includes names and dates (half of the names being companies) and provides an index to the paper files.

The combined regulatory and investigative functions of some agencies provide them with potentially a large store but, pending electronic submission or back record conversion, they remain relatively unused. At the Waterfront Commission the 'store' is potentially available for both regulatory and enforcement functions; it is based on the fact that no-one working in the Port can be registered without a criminal records check and fingerprinting. This store is now being computerised. At DOI the store is largely paper records and derives from various sources including financial disclosures that city employees above a certain salary level are obliged to complete (the Ethics Commission also receives similar disclosures) and forms on which contractors have to disclose previous dealings with the City, arrest records and so on. Some of the 'intelligence' files generated by DOI's own investigations are automated.

In the UK the computerisation of criminal and vehicle registration records has been well-established since the 1970s - the first Police National Computer (PNC) was operational in 1974. Some element of criminal 'intelligence' has always been present on some of the PNC's indexes, for example, those for *modus operandi* and Criminal Names in the sense that they might relate to allegations rather than convictions (Campbell & Connor, 1986, 226-56). Although the computerisation of criminal intelligence information started also in the 1970s it has proceeded more slowly and unevenly. In 1972 the Home Office offered to finance a pilot project in Thames Valley police for the computerisation of local intelligence records and the 'Collator Project' became operational in 1977 (see chapter four for general discussion of collators). It was not successful - Home Office evaluations found no evidence that it had had any effect on crime rates and it accumulated large quantities of unevaluated information to no apparent purpose (Campbell & Connor, 1986, 207-11).

In the 1980s the police failed for years to catch a serial rapist and killer - known as the 'Yorkshire Ripper' - partly as a consequence of poor co-ordination between different forces but mainly because of what was seen as a failure to manage and evaluate the massive information flow generated by the investigation. Thus the Home Office Large Major Enquiry System (HOLMES) was developed to manage the flow of documents and related actions in major inquiries and, as such, clearly had successes. However, although HOLMES could be used for the development of intelligence within the context of specific inquiries, it did not necessarily have utility for police beyond those inquiries and, since forces were able to acquire different specifications, did not necessarily contribute to cross-border investigations at all (Ackroyd *et al*, 1992, 150-56).

Most, but not all, forces had computerised criminal intelligence systems by 1997 but, in the tradition of force autonomy in the UK, these would not necessarily be capable of exchanging data; indeed it is common for forces to acquire intelligence systems that are not compatible even with their own records systems. For example, Merseyside Police based its Force Intelligence System on the 'Oracle' system in 1992 because it would be capable of interfacing with the systems of some other forces but in the knowledge that it was not compatible with the force's own Integrated Criminal Justice System for crime reporting and prisoner processing (Barton & Evans, 1999, 22). Similarly, in North Wales, prior to the acquisition of a new criminal records system in 1997 there was no computerised link between the criminal records and call-logging systems. Even the new records system would not be directly integrated with call-logging because of their different operating systems but they were intended to be linked by means of i-connect software (Hennings, 1997).

Hitherto, the systems operating regionally and nationally have not always been integrated but some effort is now being put into correcting this. Part of the credibility problem for NCIS when it was established in 1992 was that it was provided with a second-hand computer system from the Metropolitan Police. This was developed *via* the integration of various products into a system called ALERT in 1995. Information is passed to NCIS via Microsoft Mail and stored in Word. The store may be searched *via* a Memex Information Engine either by specific field (names, vehicles etc.) or by a 'generic' query. Output may be presented in Word, Excel or i2 visualisation (some examples of this are illustrated in chapter nine) and dissemination can be achieved by re-embedding the product in Microsoft Mail (Lander, 1998). The main sources of information will be those agencies with formal connections to NCIS or who are represented in NCIS. Therefore, the Regional Criminal Intelligence Officers, who worked with the RCS prior to the formation of the NCS, and Customs may provide data to ALERT but there is no automatic process here. Information will only go into ALERT to the extent that the information is available to someone who has the time and will to put it in.   Thus, while all police forces are potentially sources of information for the system, how much they actually provide depends largely on what they think they will get from NCIS in return which, apparently, is not a lot. Respondents to ACPO's study of inter-force crime identified NCIS as the source of just one percent of the operations they conducted (Stelfox, 1998, 403).

Hitherto Customs have used two main databases for intelligence purposes. The General Reference Information Database (GRID) is essentially their equivalent of the PNC including primarily factual records such as ships manifests. CEDRIC is the primary database used for the development of intelligence consisting not just of offender records but also of suspects and other information. It can be accessed by either a nominal or cases and is used for analytical purposes *via* Excel and i2 for charting. Customs have also developed a data warehouse, intended to become operational in 1999, aiming to provide access for investigative and intelligence purposes to a variety of different data bases. These would include all internal databases (VAT, excise, and trade statistics as well as GRID and CEDRIC), the databases of other government departments such as the Inland Revenue and Companies House as well as other law enforcement agencies and open sources. The user requirements of the resulting warehouse include free text retrieval, geographical information, data visualisation, SPSS for statistical analysis including reliability indicators and data-mining for forecasting, profiling and pattern recognition, for example, for the identity of 'fraud-prone' traders (Fuchter, 1998).

Not surprisingly the Government seeks to increase similarly the compatibility of police systems, most recently by the establishment of the Police Information Technology Organisation (PITO). This was set up in

1996 and has now been placed on a statutory basis by the Police Act 1997. Its function is to carry out research into police ICT needs and to develop and implement a co-ordinated national strategy for information systems and surveillance technologies known as the National Strategy for Police Information Systems. Projects to date have included planning for the next generation PNC, developing national databases relating to stolen property, DNA and fingerprint matching and seeking to establish standard software use including criminal intelligence systems (Uglow & Telford, 1997, 109-112). Another development seeks to overcome the heterogeneity of force systems: the Police Networked Information Project aims to provide forces with access to intelligence held in other areas and the contract was due to be awarded in July 1999. One of the bidders proposed to solve this problem by means of a Wide Area Network linking data warehouses in each force. These would be designed to handle data put in from whatever other systems the force was using (www.bull.co.uk).

Given the long tradition of force independence in the UK, the extensive ICT capital now invested in different systems and the spirited competition between technology suppliers, PITO clearly faces major challenges in pursuing the goal of standardisation. For example, Her Majesty's Inspectors of Constabulary (HMIC) noted in a recent report on Merseyside Police a range of financial, staffing, skills and management worries as to whether the force was ready to 'migrate' to the new strategy (HMIC, 1999, paras. 4.23-24). There are further reasons for questioning the extent to which integrated ICT 'solutions' can resolve problems of the intelligence process. For example, these systems still operate within a context in which forces select what intelligence they will make available to others by means of placing it in the 'data warehouse'. Undoubtedly, practitioners would all prefer to have access to others' intelligence but they will feel very differently about making their own intelligence available *via* the system. The crucial issue is control: bi-lateral exchanges of information between officers who know and trust each other are extensive and oil the working of an otherwise poorly-integrated machine but it will not run smoothly just because the technology is up-graded. Networked systems mean that forces lose control over their intelligence and this will discourage them from making available anything they consider sensitive.

Further, as things stand, there are several very good reasons for keeping ICT systems separate. Apart from the issues of privacy and data protection that may be of greatest concern to outsiders, worries about the security of sources will lead practitioners to minimise the number of people aware of certain information; this is the basis for the 'need-to-know' doctrine that is central to intelligence work but which runs counter to the pressures within law enforcement for increased sharing. Other reasons for segregating data relate to the need to distinguish 'evidence' from 'intelligence' and the disclosure rules: clearly police wish to minimise the chances of information on one operation being disclosed during the trial process following another and integrating databases increases the possibility of inadvertent disclosure. Of course, integrated systems will incorporate levels of access and security firewalls but the risks are increased.

The mechanism for information to be put into the database in North Wales was the completion of a CID56 form by officers that identifies the source (or pseudonym if an informant) and subject of the information. These are left with the Intelligence Officers in the Crime Management Unit who in-put the data into the system. A problem faced was the contradiction between rapid in-put (and therefore availability) of information, on the one hand, and quality control on the other. The argument for not having officers directly putting information in themselves was one of quality control both in relation to the information put in and the 4x4 gradings given (see Figure 7.1). The FIB trained those - mainly Local and Field Intelligence Officers at divisional level - who were allowed to in-put material. Clearly there had been problems in some areas with delays in-putting or LIO/FIOs spending too much time on what some saw as essentially copy-typing when they should be spending more time developing files. The same problem was identified in Merseyside (Barton & Evans, 1999, 26) One North Wales division had sought to ease this problem by employing a civilian to input data on the basis of the LIO's assessment of quality but in another area where there was a civilian in-putter, she had been sent to work elsewhere and the LIO was doing the job, describing her work as seventy percent administrative.

Early paper information databases tended to accumulate much information that was unevaluated in terms of relevance, difficult to access and rarely weeded of out-of-date information. A greater emphasis on targeting (see chapter six) has reduced somewhat the first problem and computerisation of the 'store' has facilitated access and weeding. In general, material and reports put into the system are allocated a 'life' and operators are alerted when that is to expire so that a decision is prompted as to whether the data will be maintained or allowed to expire (see also JUSTICE, 1998, 87-92).

Figure 7.1   4x4 Evaluation System

The 4x4 system will be used and all information reports will indicate the source, which will be evaluated in accordance with the following guidelines:

SOURCE CODE	DEFINITION
A	When there is *no doubt* of the authenticity, trustworthiness and competence of the source; or if information is supplied by an individual who, in the past, has proved to be *reliable in all instances.*
B	A source from whom information in the past has, in the *majority of instances, proved reliable.*
C	A source from whom information in the past has, in the *majority of instances, proved unreliable.*
X	In the case of *previously untried* sources where there *is doubt* about the authenticity, trustworthiness or competency.

The accuracy of the information will be graded in accordance with the following guidelines:

INFORM- ATION CODE	DEFINITION
1	When the information is known to be *true without any reservation.*
2	When information is *known personally to the source* but is not known personally to the reporting officer
3	When information is *not known personally to the source* but is *corroborated* by information already recorded
4	When information is *not known personally to the source* and it *cannot be corroborated* in any way

Information which is classified as Code 4, and that received from a Source C or X and classified as 2, will be regarded as suspect by all officers. Dissemination of such information outside the intelligence system must include a clear warning that the information is regarded as suspect.

This discussion of the store of information has concentrated on the *organisational* store, that is, information that has been passed into and kept within a formal paper or electronic system. Yet information often remains in people's heads and is not for whatever reasons - time, forgetfulness, fear, laziness, ambition - passed on. In policing, where information is the pre-eminent means to status, the tendency of officers to keep information to themselves has been well-established and represents a central challenge for attempts to establish more formal and corporate systems of intelligence (Manning & Hawkins, 1986, 145-6; Sparrow, 1994, 125-6).

Those at the centre of the new intelligence structures see this as requiring key shifts in organisation and culture. Organisationally, the main ways of seeking to encourage a greater flow of information are functional and spatial. Acknowledging the general rule that the greater the degree of organisational specialisation the less the flow of information between specialist groups, a shift towards more local and generalist teams can be observed. For example, in one North Wales division, the concentration of specialist CID teams at divisional headquarters had been reduced in order to have teams working in different areas and more use was made of joint uniformed and plain-clothes teams. Similar 'proactive' teams were introduced in Merseyside also, though in the longer term the intention was to disband the teams in favour of all uniformed and detective officers working proactively (Barton & Evans, 1999, 14-5). It remains to be seen whether this ambition would be achieved in the longer term given the regular tendency for the proactive to be subordinated to the reactive in policework.

Spatially, attempts are made to locate the CIU where least effort is required on the part of uniformed patrol officers and detectives to use it, for example, on their way to the canteen. Culturally, the objective is to shift officers' perceptions to seeing information as a commodity to be traded, not hoarded. Much CIU effort goes into encouraging the sharing of information between existing intelligence units and also encouraging other members of the law enforcement community to contribute. The problem facing CIUs in this endeavour is the value of information to the 'owner' as something with which to trade; and the ingrained reluctance among police officers to share. Several interviewees acknowledged that, historically, CIUs had the reputation among other police for merely gathering and storing information and making no useful contribution; therefore they were anxious to change this culture and would use the provision of information on suspects to patrol officers as a means of getting them to feedback further relevant information (e.g. MacPhee, 1995).

In the UK a major failure of the collator system is recognised to have been that they did not provide adequate, useful information to patrol and investigating officers. Now more effort is made in North Wales *via* instruction in the basic workings of the 4x4 system and the role of CID56s towards the end of probationer training and continuing education with posters and briefing sheets. Greater efforts are now made by Field and Local Intelligence Officers to go to patrol briefings, at least of the day shifts when they are at work, and also to visit outlying stations weekly. (See further below regarding dissemination.) Evidence from the Merseyside study (see Table 7.1) suggests that briefings by LIOs certainly have potential as an effective method of improving the flow of information:

Table 7.1 **Police views on improving the flow of information** (n=290)

	No.	%
Meetings between all ranks	66	22.8
Regular intelligence bulletins (eg from LIO)	204	70.3
Attendance on parades by management	95	32.8
Attendance on parades by LIO	205	70.7
Attendance on parades by others (eg CID, CPO)	205	70.7

Other ways of encouraging uniformed officers to provide information vary from simply flattering their efforts to having them brought into operations that result from their information, for example the execution of a search warrant. Several interviewees commented that supervisors can encourage officers to pass on information and this was reinforced in some areas by using the number of CID56s put in by a uniformed officer as a criterion for promotion to CID. In Merseyside 71% of officers said they were encouraged to provide information to the LIO or Intelligence Co-ordinator and two-thirds did so at least weekly - see Table 7.2. The main sources of this information were observations, arrests and street stops (see Table 7.3). Only just over a quarter identified informants as the source of their information; this is because 70% of officers did not have any (see chapter eight).

As well as efforts to encourage a greater flow into the CIU, attempts are made to improve the quality and utility of the information provided. The RCMP makes a number of suggestions for the key elements to be included in collection plans: membership of criminal organisations, geographic data on targets, hierarchy of the organisation (if any), criminal activities of the group, 'legitimate' business activities, financial data and connections to other criminal organisations (RCMP, 1995, 5-6). CISO trains people in the formatting of reports under the following main headings: Title, Principal(s), Information (including an evaluation of its credibility), Comments (some analysis of what the information means), Dissemination and Author (Faul, 1995).

Table 7.2   Police provision of information

How often do you provide information?	Are you encouraged to provide information?			
	Yes	No	Total	%
Daily	47	7	54	18.6
Weekly	109	25	134	46.2
Monthly	26	12	38	13.1
Less often	24	18	42	14.5
Never	-	14	14	4.8
No answer	-	-	8	2.8
Total	206	76	290	100.0

Table 7.3   Police sources of information (n=276)

	Number	%
Recent arrests	175	63.4
Registered informants	72	26.1
Public	115	41.7
Your own observations	214	77.5
Stop checks/street searches	186	67.4

Stores of information, however large or technologically sophisticated are not much good if they are not used. In the UK collators were frequently criticised for the fact that they accumulated stores of information that were used neither to develop and disseminate intelligence

nor much consulted by investigators (Kinsey, 1985, 62). Has the introduction of intelligence-led policing increased the use of the 'store' before
other methods of information gathering are considered?   Merseyside
officers were asked both about how the organisation provided them with
information on targets and crime patterns and how often they searched the
Intelligence System themselves. They confirmed the previous research that
the information police find most useful is what they receive informally from
colleagues (Manning, 1992, 364) - see Table 7.4.   Encouraging for the
proponents of the new intelligence-led structures, however, is that two-
thirds of officers also refer to information from LIOs as an important
source.   The low figure for crime tasking meetings is probably explained by
the fact that most officers do not actually attend them, and the fact that
management comes bottom in the list of sources of information re-affirms
the 'street cops' view of their limited utility (Reuss-Ianni & Ianni, 1983).

Table 7.4   How police receive information on targets (n=290)

	Number	%
Informally from colleagues	202	69.7
From the LIO	191	65.9
Parade briefings	171	59.0
Notice boards	148	51.0
Regular bulletins	126	43.4
Crime tasking meetings	46	15.9
From management	44	15.1

In other respects, the message of intelligence-led policing has some way to
go.   42.8% overall including 40% of uniformed PCs and 65% of detective
constables and sergeants, say they use the Force Intelligence System often
(see Table 7.5). However, even allowing for the fact that some jobs have a
greater need for intelligence than others, the fact that one-quarter of officers
said they rarely or never consult the system must be seen as a problem.

Open sources

Once a law enforcement agency has exhausted its own store of information, then potentially the quickest and cheapest source is information already in the public domain. Three main types are identifiable: surveillance of public space, published information and that elsewhere in the state sector. The first of these has always been central to the Anglo-Saxon model of policing:

> The prevention of crime was stressed as the first duty of the new Metropolitan Police constables, and the whole system of beat patrols - largely what the parochial nightwatchmen had been doing, often successfully, for a century or more - was ostensibly designed with this in mind. (Emsley, 1996, 25)

Table 7.5   Police use of computerised intelligence system

Rank/job	Often	Occasi-onally	Rarely	Never	Total
Uniform PC	63	54	10	30	157
DC (reactive)	20	5	1	-	26
DC (Proactive)	14	2	-	1	17
Admin. Constable	11	3	-	4	18
Uniform Sgt.	4	15	8	11	38
DS (reactive)	2	8	-	-	10
DS (proactive)	3	2	1	1	7
CMU Sgt.	2	-	1	-	3
Intelligence Manager	4	-	-	-	4
Custody Officer	1	3	-	6	10
Total	124	92	21	53	290
%	42.8	31.7	7.2	18.3	100

Information was to be obtained through uniformed patrol - there was great resistance to the Continental model of plain clothes policing until late in the nineteenth century in Britain - but, in its various forms, surveillance remains at the core of policing, both public (see chapter one) and private (Shearing & Stenning, 1981, 213).

One of the recently published codes in the UK refers to surveillance ('human or technical watching or listening') that does not involve 'interference with property' and therefore is not covered by Part III of the Police Act (that also has an associated Code published earlier in 1999). The new Code does not cover the public CCTV schemes (NCIS, 1999b, note 1A) that are now ubiquitous in towns in the UK. Since surveillance is the basic *raison d'être* of policing, it is clear that the drafters of these codes were anxious to ensure that there could be no circumstances under which police might not be permitted to conduct surveillance in *public* places. Consequently the Code asserts that 'discreet observation of the public' is a necessary part of normal law enforcement and it does not restrict police in doing that. Therefore, why bother with a Code? Because, it goes on, it relates to the 'planned deployment of covert surveillance resources' against individuals or the public at large (NCIS, 1999b, 3.1-3.2; also see JUSTICE, 1998, 30-33). The Regulation of Investigative Practices Bill distinguishes between 'directed surveillance' (targeted and covert but not intrusive), 'intrusive surveillance' (involving residences and vehicles) and covert human sources (s.25). In the new Code surveillance of *private* places where no trespass is required (if it is, then the Police Act applies) may be authorised by an Assistant Chief Constable (ACC), or a Chief Constable if knowledge of 'confidential material' may be acquired (NCIS, 1999b, 2.8-2.9). Where a target is specified then Superintendent is the authorising rank; where it is not specified or in cases of urgency, then an inspector can authorise. Although the surveillance referred to in this Code includes, where appropriate, the use of photography, there is a separate section in the Code authorising the taking of photographs of individuals or groups either to up-date police records or to obtain photographs when more extensive surveillance is not envisaged.

The relevance of published information to law enforcement intelligence is not so immediately obvious since the gathering of 'secrets', that is, information that is explicitly not public, has been suggested as the very *sine qua non* of 'intelligence' work (Shulsky, 1991, 174-7). Broader concepts of intelligence work, however, would place greater emphasis on the evaluation and analysis of information from whatever source but the general assumption that information from covert sources will be 'better' than that from open sources remains a powerful inhibition. At the level of investigative or 'tactical' intelligence this is complemented by the common sense assumption that 'criminals' are unlikely to publish their plans or intentions. Thus it is not surprising that it is those CIUs 'furthest' from the front line, in Canada for example, the RCMP's headquarters Criminal Intelligence Directorate and the former CLEU Policy Division who were the first to set up special sections for the collation of open sources and it is clearly in the area of strategic intelligence that open sources are likely to be most important. Further, they may be a great deal cheaper. For example, Robert Steele, who is literally engaged in selling the efficacy of open source 'solutions' to intelligence problems, argues that the Los Alamos National Laboratory has provided better estimates of global drug production and movements entirely from open sources at a cost of $100k compared with those of the US national intelligence community (Steele, 1995, 216).

As far as criminal intelligence practitioners are concerned, the records of other public and private agencies are a more or less 'open' source. While Privacy Acts in North America may seek to prevent the disclosure of personal records across agency barriers, in practice this tends to be surmounted in a number of ways. As we have seen, joint operations and the secondment of personnel from one agency to another give mutual access to the agencies' information stores. Other, even more informal flows of information between different law enforcement agencies are facilitated through the movement of many public sector practitioners into the private sector when they 'retire'. Another initiative aimed at facili-tating the flow of information, including between public and private sectors, is the Directory published by the Ontario Association of Chiefs of Police, giving contact personnel for all major corporations, provincial ministries and police forces:

> This directory is a working reference for police officers and security managers. Its purpose is to facilitate the exchange of information - resources, ideas and technology - that is essential to the success of both public and private protection efforts. (OACP, 1995, 1)

When it comes to police approaching other state agencies the lack of interchange of personnel is likely to increase the relevance of the relevant protocols and data protection measures. In the UK there never has been any restriction on the disclosure of aggregate information to police but that of personal information has been limited since 1984 by the passage of the Data Protection Act. However, disclosures for the purposes of 'prevention and detection of crime' were exempted from the Act and it is likely that the general willingness of staff in many state agencies to co-operate with police overcame the occasional problems.

In the UK the sharing of information between police and other state agencies, especially at the local level, has recently been the subject of encouragement by the Crime and Disorder Act 1998 that is being implemented at the same time as an up-dated Data Protection Act. Where public bodies had not previously had power to disclose information to the police they can now do so for the purposes of the Crime and Disorder Act (s.115). Since the Act places a statutory duty on local authorities to do all that they reasonably can to prevent crime and disorder in their area (s.17), there will be little in which law enforcement may be interested that may not be disclosed to them. The Act provides the power to disclose, it does not require it, but the message is clear:

> The public rightly expects that personal information known to public bodies will be properly protected. However, the public also expects the proper sharing of information, as this can be an important weapon against crime. Agencies should, therefore, seek to share information where this would be in the public interest. (Home Office, 1998, Ch.5 Annex A)

There is another very important means by which police seek information, that is, by using traditional techniques of stops, questioning, searching and interrogation. Although these methods may depend on the use of police powers, they are still 'open' in the sense that the person subjected to them knows that they are happening. Searching and interrogation (or, as UK police now prefer to describe it, 'interviewing') normally take place in the context of specific investigations and their object is 'evidence' rather than 'intelligence', though it may emerge 'collaterally'.

The police use of street stops is justified primarily in terms of crime prevention but they clearly have potential for gathering information (see Table 7.3). For example, in the UK, stop and search carried out under PACE s.1 should only be conducted when police have 'reasonable suspicion' that the person is carrying stolen or prohibited articles. Where, as for example in London, the records of these searches are stored electronically, they provide a large data base with intelligence potential. A recent study of the use of stop and search in London found that police strongly value it as a means of obtaining information but noted also that the approach to using the information in order to develop intelligence was 'at best, patchy'. This resulted from officers not understanding adequately what information should be submitted and great variation between CIUs in analysing what they had (Fitzgerald, 1999, 18-20). The study also notes that many officers regard the prime purpose of stop and search to be tracking 'known' or targeted individuals. If stops are based on what amounts to the harassment of those who have criminal records or associations, then this is certainly contrary to the Code of Practice (Home Office, 1995, paras.1.6-1.7). Research into the exercise and control of covert measures has been much more limited than into these 'open' police investigative powers, for example in Canada (Macleod & Schneidermann, 1994), UK (Bottomley *et al*, 1991) and US (Walker, 1999).

Covert technical surveillance

The research for this study focused on the role of CIUs themselves, especially the analytical process, and did not attempt any overall examination of the extent of the use of technical measures in law enforcement. In general, most information gathering is carried out by investigative teams and, depending on circumstances, these might also include an analyst (this issue is discussed further in chapter eight).

There is what seems to practitioners in other countries a curious anomaly in the UK by which any evidence as to the fact of telephone interception or material derived from it may *not* be introduced into court as evidence (IOCA, 1985, s.9). Precisely why this was included is not clear, though it is understood that the main intercepting agencies such as GCHQ would have had to transform their internal procedures if they were to meet evidential as opposed to intelligence requirements. This regime has required great care so that evidence was not rendered inadmissible in court by 'contamination' with the product of interceptions. Accordingly, the police use 'Readers' - usually Detective Inspectors, who may read and make notes on transcripts for the purposes of an investigation but which should be returned at the end for destruction. After allegations that transcripts had been leaked to a criminal, an enquiry into these procedures at NCIS was conducted during 1996 and managed to account for around 900 notebooks that had not been returned (*Independent*, January 22, 1997). The subsequent general review of NCIS procedures also involved the IOCA Commissioner who later reported that 'a more streamlined and efficient system' was to be implemented (IOCA Commissioner, 1998, paras.53-6). In that same Report the Commissioner noted a relaxation of the strict rule on the destruction of intercept material in the case of 'professional criminals' where the material could be considered to have intelligence value beyond the particular case for which it was gathered (1998, paras.20-22).

When the UK Government published its thinking in June 1999 on the need for IOCA to be brought up to date, it claimed that it had reached no conclusion on the issue as to whether the rule against admissibility should be changed. Against the obvious argument that the product of interception can be very useful as evidence, it countered that this might be only a short term gain as the exposure of interception capabilities would encourage an increase in counter-measures by those it targeted. A further concern is that, if material was to be used as evidence, on the principle of 'equality of arms' as part of the European Convention's guarantee of a right to a fair trial, then it must be available to both sides and the Government could not think of arrangements that would be 'both practicable and affordable' (Home Office, 1999, paras.8.1-8.8). This is curious - the desire not to expose techniques is not unique to interception - it applies in all areas of covert policing. In those other cases the Government relies on Public Interest Immunity (PII) certificates in order to avoid the disclosure of sensitive material but that possibility was not explored here. This is possibly because non-disclosure of evidence by the prosecution, with or without PII certificates, continues to embarrass the state by producing miscarriages of justice - the latest being the ECHR decision that the so-called 'M25 Three' were denied a fair trial for murder because the prosecution withheld key documents (*The Guardian*, April 9, 1999). Nor did the Government examine the possibility that the 'equality of arms' issue might be easier to resolve if interception were to be authorised by judges rather than ministers, a prospect that is dismissed in one brief paragraph of the consultation document (Home Office, 1999, para.7.2).

In the Regulation of Investigatory Practices Bill, the general exclusion of disclosing in evidence the fact or product of interception is retained (s.16) though there are exceptions so that cases involving breaches of the law relating to interception itself can be pursued and so that prosecutors and judges may be told of interceptions. Judges may then, in exceptional circumstances, order that the disclosure be made to the defence in the interests of justice (s.17). Thus it is clear that the Government envisages no general use of interception material in trials.

When the UK Government passed IOCA in order to achieve minimal compliance with the ECHR *Malone* decision, it did not take advantage of the opportunity to provide a systematic regime for all forms of intrusive technical surveillance, especially 'bugging'. Instead it issued guidelines that required the authority of the chief constable, that the investigation concerned serious crime, that normal methods of investigation must have been tried and failed (or be unlikely to succeed) and there must be good reason to believe that use would lead to arrest and conviction or prevention of an act of terrorism (Home Office, 1984). Once the security intelligence agencies were required to obtain warrants from a minister for 'interferences with property', there was judicial criticism of the anomalous position of police and the Home Office sought to regularise the situation in the 1997 Police Act.

But this Act still enabled the police normally to authorise their own operations. After the event both police and security applications are subject to review by a judicial commissioner. The Police Act applies to Customs as well as police: 'normal' authorisations can be given by chief officer ranks but must be immediately notified to a commissioner who may quash it if s/he believes there were no reasonable grounds for the authorisation. Where 'sensitive information or locations' are involved then a commissioner's express approval is required before an authorisation can take effect, although in cases of urgency this requirement can be waived. Sensitive information includes matters subject to legal privilege and confidential journalistic material; the locations covered are private dwellings, hotel bedrooms and offices (Uglow & Telford, 1997, 77-8). This provision for prior approval was inserted into the Act only after vociferous opposition to the Bill that proposed only *post hoc* review. The new Act has required some modification to force procedures, for example, Merseyside developed an 'intrusive surveillance management unit' within the FIB that was to become the 'gatekeeper of legality and ethics' regarding surveillance with the role of managing PII issues and the tasking of the dedicated surveillance unit (HMIC, 1999, paras.4.21-22).

One of the new Codes issued in UK in 1999 also deals with the interception of communications and accessing communications data (NCIS, 1999d). The first two parts tell us nothing new since they cover the authorisation of the interception of mail, public telephones and wireless telegraphy that are already provided for by the IOCA 1985 and the Wireless Telegraphy Act 1949. Such authorisations are available only in the cases of national security, serious crime and the safeguarding of 'economic well-being'. The code re-states the provisions of IOCA whereby warrants must be obtained from a minister and that complaints about the use of the powers should be addressed to the Tribunal. The section on wireless telegraphy covers the interception of private radio systems including pagers. As 'crown servants' Customs officers are already empowered under the 1949 Act to intercept these but the Code states that these will now be authorised by an Assistant Chief Investigation Officer while the police - who are not crown servants - will require authorisations from the designated official in, for example, the Home Office, who acts under ministerial authority (NCIS, 1999d, 3.9). Where one party to a communication consents to the interception then a ministerial warrant is not required and an ACC can authorise. JUSTICE specifically argued in their report on covert policing that interception should be subject to the same controls whether or not one party consents (1998, 14-30).

This Code and the Regulation of Investigatory Practices Bill (ss.20-24) bring into the authorisation regime the issue of police obtaining communications data from system operators. Much lower authorisation thresholds and a wider array of circumstances apply to this compared with interception. A request for information identifying a particular subscriber can be authorised by an Inspector. A Superintendent's authorisation is required for access to data other than this, for example, the source and destination of all calls to or from any particular number - 'metering'. The police make regular use of such data for investigative and intelligence purposes, for example, in charting telephone networks. IOCA 1985 made no provisions for the authorisation of this technique - the only reference was in one of its Schedules absolving the operators from liability in disclosing such information for the purposes of the prevention and detection of crime and in protection of national security. Consequently the practice has developed through routine dealings between operators and law enforcement agencies.

At local level the cost and practicalities of some these techniques is prohibitive of their routine deployment. Certainly, interception is used relatively rarely while 'metering' information is obtained more regularly. Similarly, specialist equipment for forms of surveillance not requiring 'entry on or interference with property', for example cameras, cars, radios is relatively easily available from the Technical Support Unit (TSU). Until recently these had been located regionally in the UK, now each force is developing its own. But where the placing of bugs or cameras requires covert access to property then both the length of the authorisation procedures and the costs of the operation increase considerably.

The law and regulation tend to lag behind technological innovations, especially in areas that expand as rapidly as ICT. This was not a particular problem for state agencies while they were at the cutting edge of developing these technologies, for example, satellite communications, and their use and interception of these media could remain relatively unchecked until the law caught up. But a combination of technological developments and privat-isation has removed state agencies from their privileged position and left them as 'just another' player in transnational markets. The private sector is now the main provider of international communications systems and the rapid growth of the Internet and cellular phone networks has seen states having to play 'catch-up'. In doing so they face a specific case of the contradiction that besets all the crime control efforts of the developed 'neo-liberal' democracies, that is, facilitating the maximum freedom for markets for 'legal' goods and services, specifically electronic commerce, while restricting the use of these networks by money-launderers, child pornographers and other 'undesirables'.

Both interception *per se* and the advance in the encryption of messages pose problems for states. The precise extent to which intelligence agencies are now intercepting law enforcement targets remains secret but it is certainly happening - in the UK the process was legalised by the Intelligence Services Act adding 'serious crime' to the national security mandate of GCHQ. Regarding decryption, the initial preference among governments both in North America and the UK was for a 'key escrow' system in which an independent authority would be established as a 'trusted third party' who would possess the 'key' necessary to de-code any communication but who would make that available to state authorities only on production of a warrant. This proposal ran into major opposition on a variety of grounds, for example, its practicability and the extent to which any authority could earn the required trust among service providers or, from a different point of view, whether law enforcement agencies could gain access as readily as they would like (*The Times*, May 27, 1999). Different countries clearly had very different views regarding the way forward, for example in Canada the law enforcement lobby was unable to prevent a liberal policy in relation to electronic commerce (CASIS, 1998, 19) and France sought to resist US pressure by declaring unlimited public use of encryption (*Intelligence*, N.95, March 22, 1999, 15). In 1997 an EU Report rejected the US preference for key escrow and sought to maintain Europe at the forefront of electronic commerce by encouraging uniform and effective encryption standards (*The Guardian*, October 16, 1997).

Yet, since 1993 the FBI has led an attempt to provide a co-ordinated approach known as the International Law Enforcement Telecommunications Seminar (ILETS) in which International Requirements for Interception have been developed. As well as EU countries and the US this involved Canada, Australia, New Zealand and Hong Kong. The proposal became EU policy in 1995 and requires manufacturers, Internet Service Providers and network operators to build in 'interception interfaces' to the Internet and all future digital communications systems (*The Guardian*, April 30, 1999). These would be available to law enforcement personnel subject to the whatever specific legal procedures were legislated in each country; in some, for example, the US, Canada and Netherlands local legislation has already been enacted. In the UK Part III of the Regulation of Investigatory Practices Bill empowers police and security agencies to issue notices to Internet service providers obliging them to disclose codes and creates a specific offence of 'tipping off' if someone discloses the existence of such a notice (s.50).

These efforts of the law enforcement community to catch up with the Internet mirror the already well-established interception methods deployed by security intelligence networks. The global intelligence network run under the UKUSA agreement since the 1940s (Richelson & Ball, 1990) was developed into the ECHELON system during the 1970s. Based on the interception of satellite, cable and Internet communications, intelligence customers can task computers at each collection site. Targeting is carried out by local 'dictionary' computers that store whatever names, topics, addresses, telephone numbers and so on are selected as of interest.   The sorting and selection process is similar to that carried out by search engines on the Internet.   Information regarding this system has only recently emerged in a report to the European Parliament's Science and Technology Options Assessment Panel based on research by Duncan Campbell, the investigative journalist (Campbell, 1999; see also www.ecelonwatch.org).

Traditionally, within the area of communications intelligence a primary distinction between 'foreign' and 'law enforcement' intelligence is that the latter is most likely to be targeted on specific individuals while the former, operating under general warrants, makes greater use of 'trawling' techniques. The existence of ILETS has only emerged in 1999 as a result of a leaked document and conceals, it is argued, the fact that the main US agenda throughout has been national security not law enforcement intelligence, the latter being a 'smokescreen' for the former (Campbell, 1999, paras.84-89).   Now that security intelligence agencies *are* increasingly involved in law enforcement, the idea of one of these being used as a smokescreen for the other is perhaps slightly missing the point. The networks and mixing of mandates between intelligence and law enforcement agencies are now extensive, though there may still be regular turf wars between them. Accountability in the latter was always seen to be clearer largely because it aimed to invoke the relatively transparent criminal justice process. But the increasing use by law enforcement of security intelligence methods including 'disruption' show that the two areas converge and, consequently, it becomes virtually impossible to maintain the traditional distinction that Duncan Campbell argues is vital to civil liberty (1999, para.89).

Conclusion

The new technologies for gathering, storing and processing information are the crux of both the hopes of practitioners and the fears of critics regarding the relations of state and citizens. It is clearly the case that law enforcement agencies - like any other organisation - are able to store, process and access information much more readily. But the technologies alone only provide the potential for greater impact, whether that is defined as 'crime-control' or repression. For example, they are a necessary but not sufficient factor in solving the problem of information overload. Technical means of information gathering have similarly developed rapidly though their availability is always mediated by costs, in terms both of acquiring equipment and having the resources to deploy it covertly. Legal standards to govern the use of these technologies have developed more slowly in the UK than in North America though, currently, the Regulation of Investigatory Practices Bill seeks to put into place a statutory scheme for the authorisation and review of all intrusive information gathering.

This Bill reflects the concern of state authorities with the 'security deficit' they see emerging because technological advances out-strip their ability to intercept and understand communications, but there must be concern also at the large 'democratic deficit'. In the UK, in particular, the new procedures enshrined in the Regulation of Investigatory Powers Bill are primarily concerned with the effective management of the process of authorising the range of covert techniques. This continues the trend started in IOCA 1985 which established a Commissioner to carry out a *post hoc* 'judicial review' of the process by which interceptions were authorised and a Tribunal to receive and investigate complaints. The narrow remit of these bodies has prevented them from carrying out any substantial enquiry into state practices (Gill, 1994, 290-6) but, having passed muster before the European Court of Human Rights as meeting the minimal requirements of the Convention, the model has been extended through a series of statutes to the intelligence services and police surveillance practices. The Regulation of Investigatory Practices Bill continues with this trusted model: it replaces the IOCA Commissioner with a new Interception of Communications Commissioner who will have additional powers to review the acquisition of communications data (ss. 20-24) and brings into line similarly the role of the Surveillance Commissioners (see s.55) who were established under the Police Act 1997. Another new Commissioner is set up who has the potential to review the use of human sources (s.53[3]), to which we turn in the next chapter.

8 Informants and Undercover Police

Introduction

The investigative work of all law enforcement agencies relies to a great extent on informants. Detective work has always centred around their cultivation but informants are recruited and paid in by many law enforcement agencies. For example, although Customs does not have the general law enforcement mandate of the police and, to that extent, their opportunities to recruit informants are fewer, their extensive regulatory and inspection powers with respect to traders provide for plenty of recruitment opportunities. Inland Revenue in the UK has been authorised to pay informants since 1890 and normally pays amounts anywhere from £100 to £20,000 depending on the amount of tax subsequently collected (*Observer*, December 1, 1996, 2). Some areas of law enforcement, it is fair to say, simply could not exist without them: 'Without a network of informants - usually civilians, sometimes police - narcotics police cannot operate' (Skolnick, 1994, 117).

Yet, although, compared with technical means of surveillance, 'human sources' have been used by police for much longer, are more extensively deployed, more productive and less controlled, the amount of ink spilt and statutes and codes promulgated to regulate the former is far greater. The paradox here is more apparent than real: it is the very centrality of informants in policework that accounts for the marked reluctance on the part of ministers to seek to lay down rules for their use and there is an acknowledgment within law enforcement circles that it would be impossible to devise codes that would be both consistent with law *and* enforceable without reducing dramatically the flow of information.

Relationships of power and exchange lie at the heart of law enforcement's reliance on informants:

> The informant-informed relationship is a matter of exchange in which each party seeks to gain something from the other in return for certain desired commodities. (Skolnick, 1994, 120)

What police want is information that can assist in the fulfilment of organisational and personal goals while the informant may want relief from a legal sanction or a guilty conscience or to receive money (or all three). But the power resources available to the 'negotiators' of this exchange are by no means equal - those of the police are reinforced by formal legal authority while those of the informant are compromised by the possibility that authoritative sanctions may be imposed at any time and the fear of exposure to associates.

Different literatures use the terms 'informant' and 'informer'. The latter may be interpreted more specifically as the giving of information or making of accusations *against a specific person* while the former refers more generally to the provision of information (for example, regarding locations or activities as well as about people). However, the difference is not significant and the more general term will be used here. What is more important is to distinguish different types of the phenomenon.

The first type, we might call them 'sources', are distinguished by the fact that they are unpaid and expect no specific benefit from police in return for their information. These people may range from those who simply want to be a 'good citizen' by passing on their observations of suspicious activities to those whose jobs give them more regular access to information and who might be cultivated by police, for example, hotel receptionists, postal workers (e.g. Skolnick, 1994, 120). Exceptionally, these people may have to be given protection as potential witnesses if their safety is under threat.

Second, there is the 'informant' who is distinguished by the fact of receiving some reward and is increasingly likely to be 'tasked' by police in an attempt to increase 'productivity'. The police definition in the UK is now:

> an individual whose very existence and identity the law enforcement agencies judge it essential to keep confidential and who is giving information about crime or about persons associated with criminal activity or public disorder. Such an individual will typically have a criminal history, habits or associates, and will be giving the information freely whether or not in the expectation of a reward, financial or otherwise. (NCIS, 1999a, 1.14.1)

Third, there is the 'participating informant':

> an informant who is, with the approval of a designated authorising officer, permitted to participate in a crime which others already intend to commit. (NCIS, 1999a, 1.14.2)

What distinguishes this person from the second type is that, in order to maintain their personal safety and usefulness to police, they are permitted by their handlers to participate in some continuing criminal enterprise.

Fourth, there is the 'supergrass' - these are likely to have been participating informants but they agree to testify in court against their former associates, usually in return for a significant sentence discount. ACPO's official term for these is 'resident informant'. Once they have appeared in court, of course, they will be of no further use as informants and may well be in danger so they and their families will require expensive re-location with new identities. It has been reported that over 700 informants and their families were dealt with in this way between 1978 and 1995, in some cases involving re-location abroad (JUSTICE, 1998, 51. Morton, 1995, xiii; JUSTICE, 1998, 41; Innes, 2000, 359-62 offer slightly different categorisations. For a discussion of informants in the related context of security intelligence see Gill, 1994, 154-61).

At different times, the same person might be any of these types or, over time, might be 'developed' by police from one to another, or, most dangerously, might simultaneously be different types for different agencies. The development of centralised informant registers within forces and, in some cases, nationally as at NCIS, is supposed to prevent this as well as enhancing more effective targeting of informants but, as is common in this area, the relationship of practice to policy may be weak. The UK Police Inspectorate noted that there was a general lack of awareness and therefore under-use of the NCIS register (HMIC, 1997b, 17).

Authorisation and guidelines

As we saw in the previous chapter, prior judicial authorisation of covert technical surveillance only came about in the UK because of the extensive criticism of the government's original proposal to give statutory expression to the then practice of police authorising themselves. Ironically, the parliamentary criticism did not extend to the absence from the legislation of *any* regulatory framework for what is arguably the far more intrusive, frequent and problematic use of informants. This mirrored the long-standing official reluctance to regulate these practices and only in recent years has the situation started to change. In Belgium (Van Outrive & Cappelle, 1995) and Netherlands (Klerks, 1995), for example, prosecuting magistrates now have a role in the approval and supervision of informants. In the Netherlands this was introduced as a result of a parliamentary inquiry into the way in which participating informants generated their own highly lucrative drugs importations (de Roos, 1998, 95-110; Field, 1998, 378-9).

Certainly, the practical and ethical issues involved are significant, though they are now receiving more academic and official attention in the UK also. In part this is because the police there are, for the first time, having to contemplate the human rights implications of their policies and practices. Therefore police managers are trying to enact procedures that will protect the police both from legal challenges and the scandals that regularly erupt around the relationships between police and informants.

Police themselves will often, usually privately, express great ambivalence about the use of informants: they are seen as indispensable to gathering information about crime yet they are capable of causing great grief. The first Home Office guidelines on the subject were issued in 1969 only after stinging criticism of the police by Court of Appeal in *R. v Macro* (Robertson, 1976, 183-8) but even subsequent up-dates and further ACPO or individual force guidelines concentrated on procedures for handling informants and did not cover the vexed issue of recruitment or other ethical issues. For example, the (unpublished) Guidelines issued by Merseyside Police in December 1994 were fourteen pages long and dealt with handling, supervision and payment but said nothing about recruitment.

Another reason for the greater attention now paid to this issue is the policy shift towards intelligence-led policing in general and the greater use of informants in particular (Audit Commission, 1993). This pressure has not been felt evenly throughout the police, for example, at national and regional level where the focus is on relatively small numbers of major criminal activities, informants have always been an important part of policing and technical means of surveillance are often as important. At local level, detective policing has traditionally made systematic use of informants but now the emphasis is on recruitment and targeting by both detectives *and* uniformed officers. Technical means of surveillance are less available locally because of resource constraints and unattractive given the long authorisation processes necessary; informants are cheaper and easier to use.

The new systems and structures developed in the mid-1990s for the recruitment, handling and tasking of informants in Merseyside has been described as almost 'a revolution'. In one area an Informant Tasking Team of two officers- one uniformed, one CID - chosen for their local knowledge and extensive use of informants, sought to collate information gathered before it went to the FIB. They also liaised with the FIB to identify informants elsewhere with information on local targets so that handlers could task them for further information. This structure was recommended for adoption through-out the force (Barton & Evans, 1999, 38-9).

In North Wales the policy announced for criminal intelligence in 1996 included:

> 3.1 Responsibility rests with *every member of the force* to actively seek to recruit informants...
> 3.4 *Superintendent, Training*, is responsible for providing staff with the skills necessary to effectively handle and control informants.' (North Wales Police, 1996, emphases in original)

While it was clear that divisional CID chiefs were trying to encourage the first of these, the training requirement was absent. Clearly, detectives are still more likely to be running informants than uniformed officers. All that a constable might receive if s/he asked to register an informant was some 'on-the-job' training from an officer with some experience; one interviewee acknowledged that this rush to recruit without appropriate training was 'dangerous'.

Barton and Evans also noted problems regarding the lack of direction for officers in handling informants (1999, 38-9) They comment that this is consistent with the traditional notion among detectives that experience is more important than formal training (1999, 29; cf. also Greer and South, 1998, 34-5) but careful examination of their data showed an extensive lack of training and, indeed, a significant mismatch between those who have been trained and those handling informants. Table 8.1 shows that, several years after the Audit Commission's recommendation to police to make greater use of informants, almost 70% of the officers in this survey were not running any and 78% had had no training. While over one-third of those who had been trained were handling no registered informants, one-fifth of those who were untrained did have registered informants! Without comp-arable figures for the numbers of officers with informants in, say, 1993, it is not possible to gauge precisely how much greater Merseyside's deployment of informants had become but this data does raise a question as to the extent of the 'revolution'. Of course, these figures do relate to registered informants and, bearing in mind Dunnighan and Norris's (1996) research, it is likely that more officers than appear here are actually running informants.

Table 8.1   Police informants

Merseyside Police Officers trained in recruitment and management of informants and numbers of registered informants, if any.

No. of registered informants	Year of Training, if any						Total	%
	None	Pre-1994	'94	'95	'96	'97*		
none	178	4	3	5	12	0	202	69.7
one	16	2	0	1	6	1	26	9.0
two	16	0	0	1	4	2	23	7.9
three	6	3	1	2	7	0	19	6.5
four	3	0	3	0	0	0	6	2.1
five & over	6	1	0	2	1	3	13	4.5
n/a	1	-	-	-	-	-	1	0.3
Total	226	10	7	11	30	6	290	
%	77.9	3.5	2.4	3.8	10.3	2.1		100

* Data collected January-March 1997 and so figures not for whole year.

It is clear that forces have taken up the suggestion to make greater use of informants but whether they have provided the necessary infrastructure of supervision and training for a controversial area of policing is another matter. The main sections of the 1999 UK Code specify the requirements for registration and use of informants, participating informants and juvenile informants. There is an escalating ladder of authorisation depending on the degrees of seriousness of crime and intrusions on privacy. So, an informant likely to be of value in the prevention and detection of crime, maintenance of public order or community safety may be registered and used on the authority of a Superintendent or of a chief constable if the informant is likely to acquire knowledge of confidential material. Assistant Chief Constables must review the authorisations given by junior officers of the registration and use of juveniles (NCIS, 1999a, 4.5) and also the use of participating informants whose participation must be 'of substantial value to an investigation concerning *serious* crime, a *significant* threat to public order or a *significant* threat to community safety' (NCIS, 1999a, 3.1, emphasis added). But the impact of these narrower terms is more apparent than real. While 'serious crime' is a higher threshold than 'crime', the definition of the other two terms in the Code fall short of specifically criminal behaviour; for example, 'community safety'

> ...includes criminal or anti-social behaviour which is intended or likely to spread the fear of crime or violence or which is intended or likely to corrupt or undermine the health and well-being of the young or other vulnerable sections of the community. (NCIS, 1999a, 1.14.8)

Thus it is difficult to imagine any set of circumstances in which police might intervene in which they would *not* be able to use a participating informant. We should not be surprised that the agencies were only prepared to issue Codes that would have this effect but it does make even more urgent the need for some external review of the actual use of informants, participating or otherwise.

The Code sets maximum periods for which registration may last (NCIS, 1999a, 2.12-14) and what the confidential records relating to contacts and assessments of the effectiveness of informants should contain (2.17). Higher standards regarding proportionality and necessity are to be applied in the case of juveniles and their use may only be authorised for a month at a time (4.1-6).

Part II of the Regulation of Investigatory Practices Bill sets out the first statutory framework for the use of informants and undercover officers under the general heading of 'covert human intelligence sources':

> ...a person is a covert human intelligence source if -
> a)   he establishes or maintains a personal or other relationship with a person for the covert purpose of facilitating the doing of anything falling within paragraph b) or c);
> b)   he covertly uses such a relationship to obtain information or to provide access to any information to another person; or
> c)   he covertly discloses information obtained by the use of such a relationship, or as a consequence of the existence of such a relationship. (s.25[7])

As with other intrusive methods, their use can be authorised if 'necessary' and 'proportionate' with the additional requirement that there be satisfactory arrangements in place regarding handlers, supervisors and record-keeping (s.28). The Bill legalises the existing distinction whereby informants will be used much more frequently and with less control than technical surveillance: informants can be authorised by any 'designated' officer compared with 'intrusive surveillance' that requires internal authorisation from an assistant chief constable (ss.31-32) and may require prior approval from the Surveillance Commissioner. There is an inherent contradiction here: since intrusive surveillance is defined to include 'the presence of an individual' in a residence or private vehicle (s.25[3]), it looks as though using informants - who are by their nature about as 'intrusive' on individual privacy as it is possible to get - will enable agencies to avoid this higher authorisation level.

Recruitment

The important ethical issues surrounding the recruitment of informers are barely addressed in the new Code. It claims that informants 'will be giving the information freely' (NCIS, 1999a, 1.14.1) and one LIO commented, that 'one volunteer is worth ten conscripts'. But most informants are recruited when they are prisoners; of Barton and Evans' sample of 100 drawn from the Merseyside Police source registry, 54 were prisoners when recruited while Dunnighan and Norris found that 84% of officers' registered informants were either under arrest or active investigation at the time of their recruitment (1996, 3; also Innes, 2000, 368). In New York similarly, the recruitment of informants is mainly *via* arrests - police believe the best time to 'flip' them is immediately before their 'disappearance' or court appearance makes their acquaintances suspicious that they have been turned. Whether or not agencies try to 'flip' people depends on a number of factors such as their willingness to co-operate and the quality of information they can provide. The imbalance of power inherent in being in police custody and the police use of psychological coercion, deceit and manipulation significantly qualifies the 'freedom' with which people may give information (Dunnighan & Norris, 1996, 3-9, Greer & South, 1998, 32-3).

Similarly, the use of convicted prisoners as informants raises serious issues. For some time these have been used by UK police primarily as a means to quickly enhance otherwise dismal crime clear-up rates.

Typically, police would visit local offenders in prison with lists of 'unsolved' local property crimes and ask if they had committed them. With the aid of small gifts of tobacco or vague hints of future favours the prisoner might well admit to doing so and large numbers of clearances could be obtained. Occasionally the use of this transparent method of cooking the books raised sufficient controversy that HMIC started to require police to distinguish between 'primary' and 'secondary' clearances in their statistical returns. However, the practice has not ceased entirely and prison-visiting itself is still encouraged by the Police Inspectorate if it is done with a view to developing informants. Thames Valley Police instituted a central team that aims to visit all prisoners convicted of burglary and car theft with a view to cultivating them as informants and also targeting prolific offenders on release (HMIC, 1997b, 17).

Barton and Evans' data provide an interesting socio-economic snapshot of their sample (this included no juveniles, suggesting that their registrations are maintained separately). The average age of the 76 men was 32, and of the 24 women was 31. Only 19 were employed and most were single. Seventy had previous convictions, the men mainly for theft (32) or drugs (12), the women for theft and handling (11). The information most frequently provided concerned drug-related offences (74), handling (53), burglary (40), 'organised crime' (17) and robbery (9). As we have seen, most were recruited while prisoners; the method of recruitment recorded for the others was: volunteers (27), victims (5), social contact (4), 'other' (10 including 2 stop-checks).

The silence in the codes regarding recruitment is a major problem since there are clearly major ethical issues if people are being manipulated or pressurised into potentially dangerous relationships with the police. These issues are raised in even sharper relief in the case of juveniles. As we have seen, formally, decisions to use them must be reviewed by an ACC but the active policy of targeting and rewarding young people with clothes, bikes, cash or whatever might well result in police pressure on them to continue with criminal associations that they might otherwise abandon (e.g. JUSTICE, 1998, 42-3).

Recruiting informants is as old as policing itself but publicly advertising for them is a more recent innovation. In recent years in the UK the *Crimestoppers* programme, funded by the private sector has established advertising campaigns asking people to ring the police and provide information about any crime confidentially in return for a reward that will be paid if the information leads to a conviction. This innovation straddles between the 'source' and the 'informant' in our typology: if such a payment is a one-off then it is nearer the 'good citizen' but if it is used by police to bolster the payments from police funds to someone who has been tasked then s/he is a traditional informant. Police clearly believe this to be an effective innovation in a number of ways. For example, the scheme has greater credibility with magistrates than the 'anonymous sources' upon which they might otherwise rely when seeking a search warrant. A 'health warning' needs to be attached to the figures in Table 8.2 since there has been no independent evaluation of the precise contribution to arrests of the information provided compared with other investigations.

Table 8.2   Merseyside Police, 'Crimestoppers' 1994-96

	1994	**1995**	**1996**
Number of calls recorded	1000	1,200	1,270
Number of arrests made	47	80	87
Value of property recovered (excl. drugs	£72,530	£137,510	£312,230

Source: Barton & Evans, 1999, 41.

Another issue on which there is silence in the Code is the use by police of non-registered informants. The new Code makes clear that informants are to be seen as corporate resources rather than for the exclusive use of their handler (NCIS, 1999a, 1.9) but this entirely proper view may collide with what can be intense personal relationships between informants and handlers. There are many factors to be weighed up by police officers in deciding whether to register their sources. For example, some sources will be happy with a personal relation with a specific officer but will be far less happy at becoming a 'corporate asset' since they may feel it increases their risks of exposure. In any event research has shown that many officers - 55% in Dunnighan and Norris's research - still run unregistered informants (JUSTICE, 1998, 43).

Management

Apart from the imminence of the Human Rights Act, another precipitating factor for increased attempts to regularise the use of informants in the UK has been continuing scandals. For example, there was major controversy around the criminal activities of police informants Eaton Green and Delroy Denton which culminated in Denton's rape and murder of Marcia Lawes in 1994. Amid a police panic around the activities of 'yardies', Green, having jumped bail in Jamaica, was recruited as an informant after arrest in London in 1991 and continued to deal in drugs and use firearms while his handler - a constable - had no supervision. Arrested along with others for a robbery in Nottingham in May 1993, Green's role as an informant was not disclosed by the Met. to Nottingham Police or the Crown Prosecution Service (CPS). When it emerged in open court the trial was abandoned and then resumed with Green as a prosecution witness. After the conviction of the robbers, Green made further confessions to London detectives that were at variance with his evidence at the trial and when, eventually, this was disclosed, the Court of Appeal quashed the convictions. Denton was known by police and immigration to have used extreme violence in Jamaica and entered the UK illegally but, in return for becoming an informant, he was released while his application for asylum was considered (*The Guardian*, February 16, 1999, 6).

These events were first exposed by the media in February 1997 (e.g. *The Guardian*, February 3, 1997) and after the quashing of the convictions for the Nottingham robbery in May 1997 several inquiries were established. One working party included representatives from the Police Complaints Authority, Metropolitan Police, National Crime Squad and members of police community consultative groups. Its report made suggestions for minimal national standards to be established in line with the ECHR and for assessments of informants' value to include issues of public confidence and offence seriousness (*The Guardian*, July 7, 1998, 11). A police inquiry was also conducted by Sir John Hoddinott, chief constable of Hampshire, who reported to the Police Complaints Authority (PCA) in October 1998 and essentially confirmed the media reports of chaotic management and law-breaking by detectives. This report was also sent to the CPS but they concluded in July 1999 that there was insufficient evidence to prosecute any police officers (CPS, Press Release 128/99, July 15, 1999; *The Guardian*, February 16, 1999, 6). Further scandal broke over the heads of NCIS and NCS in February 1999 when it was announced that their directors-general had jointly appointed an inquiry to be conducted by Greater Manchester Police after the head of the NCIS North East Region and an officer from the Northern Office of the NCS had been returned to their home forces over an issue of relations with an informant (*The Guardian*, February 27, 1999, 1).

It is not surprising therefore, that agencies will go to considerable lengths to keep informants under control. In New York, for example, a detective will place a recorder (that cannot be switched off) on the informant before they meet with a target and take it off again afterwards. Anyone going out for a 'buy' or a 'payoff' will be searched before and after the meeting and agencies will try to ensure that meetings take place under surveillance. The issue of informants has proved just as controversial in Canada (e.g. Beare, 1996, 190-6; Palango, 1994, 254-69, 279-87) and both the Metro Toronto Police and RCMP have force-wide units that attempt to oversee the registration and management of informants.

In North Wales there should be co-handlers for each informant and formally both should attend any meetings. Logs of meetings kept by the DI will include the information provided and payments made but before CID56s are entered into the computer system the information can be sanitised by the handler to protect the informant's identity. Yet, in practice the utility of co-handlers was seen as increasing the chance that, given shift and leave demands, *someone* would actually be available for a meeting rather than that both would actually attend. This reinforces other findings of a wide divergence of practice between forces on co-handling (Greer and South, 1998, 36-8).

Indeed, sound management policies and legally watertight codes of practice do not translate easily, if at all, into practice. A number of factors contribute to the difficulties of management and control of informants: police officers are reluctant to register someone until they have proved their reliability and try to minimise the risks of untrustworthy informants by keeping their dealings secret from superiors. Of course, it is precisely this which breaches guidelines and lays officers open to potential discipline. Given the pressures on police to run informants, for example, its use as a criterion for assignment to CID, the status it brings and the pressure on clear-up rates, many 'street cops' will express cynicism at management attempts to legislate morality. On the other hand, Dunnighan and Norris conclude that there is not simply a clash between 'street' and 'management' cops as to how the job has to be done; but, rather, they found a coincidence of view - over half of their sample - that it is not possible to run informants according to the letter of the guidelines (1996, 9-22). This area of policing is a particularly sharp example of the general rule that guidelines should be seen as describing a desirable yet unachievable position. Their primary function is presentational, that is, as a set of rules on which the organisation can rely if and when it decides to take formal disciplinary action against an officer or in which practices otherwise become public.

In recent years the rate at which rules regarding informants have changed has accelerated in the light of concerns about both effectiveness and propriety. The performance pressure has been manifest in clearer efforts to give informants specific targets rather than relying on information as and when informants chose to provide it. But the most significant factor affecting the effectiveness of informants is that they are protected from exposure - this clearly is crucial both for individuals and any system based on informing. Consequently police report both moral and pragmatic reasons for regarding secrecy as to identities as sacrosanct. This has particular significance for the integrity of the criminal justice process as a whole because of the controversial new disclosure regime established by the Criminal Procedure and Investigations Act 1996 (CPIA) in which the police make the crucial decisions as to whether evidence is 'relevant' and therefore must be disclosed to the defence (Leng & Taylor, 1996, 11-33). Even before this Act increased the control of the police over disclosure, Dunnighan and Norris found that non-disclosure of informants was the rule: from their sample of 114 cases involving informants they examined a sub-sample of 31 prosecution case files in which official payments were made 'but in not a single case was the role of the informant disclosed to the Crown Prosecution Service' (1996, 18). Skolnick's research in the US found similarly that narcotics police would, if possible, avoid revealing the use of informants - despite the ubiquity of informants in drugs policing, police files in less than 10% of the cases he examined revealed the use of an informant. However, that study was in the 1960s since when there have been significant judicial efforts in the US to improve the propriety of policing. Formally, now, in New York, for example, deals made with informants, for example, their need to inform handlers regarding any commission of crimes, will be in sealed records with judges.

It would be nice to think that this police reluctance to disclose would have been reduced when the new disclosure regime came into effect in the UK in 1997 but given the overwhelming concern with secrecy this seems unlikely. Indeed, other events will have served to reinforce police secrecy, for example, the revelation in October 1998 that a CPS official had been charged with conspiracy to pervert the course of justice by allegedly leaking a list of 33 police informants to outsiders assumed to have criminal connections (*The Guardian*, October 12, 1998, 1). Surveys by the Law Society and the Criminal Bar Association found 'a widespread and fundamental series of failures' by police and prosecutors to comply with the Act and an admission by the Director of Public Prosecutions himself that the disclosure rules were not being followed in 'a significant number of cases' (*The Guardian*, July 15, 1999, 8).

The extreme reluctance to disclose the presence of informants in investigations is explained partly by the fact that, if it is, then it will raise issues of police propriety that may ultimately threaten the prosecution's case. The issue of what informants received in return for their information becomes central to the integrity of the prosecution. Barton and Evans study showed the informants' motives recorded by police as financial (69), revenge (19), reform (14), gratitude (13), 'self-importance' (10), fear (3) and social conscience (5) (cf. Also Dunnighan & Norris, 1996, 3-5). A similar range of motives has been found in the US (Skolnick, 1994, 119). Yet money may not be the main motive for informants. When initial recruitment takes place an important factor is the expectation of some concession from police over charge, sentence or bail. While it is possible in the UK for the police simply not to charge someone, formal offers of immunity from prosecution may only be made by the CPS and, of course, police have no direct influence over sentencing. However, they may prepare a document (a 'text') that the prosecutor can present to the judge - secretly if required - giving details of the informant's assistance (JUSTICE, 1998, 45-8). Similarly, in the US Skolnick found that

> the coin of the informer's realm is primarily not money, but rather is to be found in the discretion residing in the position of police officer and that of district attorney. (1994, 122)

Being derived from police records, it is likely that Barton and Evans' figures over-estimate the significance of money because of the reluctance of police to record officially reasons for recruitment resting on more controversial uses of discretion such as holding out promises of bail and not charging with offences. This full range of discretionary benefits that police might use is glossed over in the UK Code of Practice that specifies only formal texts in its definition of non-cash rewards (NCIS, 1999a, 1.14.5).

Yet money is important as a regular motivation for informants, though the amounts paid routinely are not large. For example, in 1992 the average paid by forces was £19k with an average individual payment of £100 though in specific cases much higher amounts might be paid (JUSTICE, 1998, 45). But since 1992, largely as part of the new emphasis on informants, the amounts available have increased considerably, for example, in Merseyside payments in 1995-5 were £60k rising to £80k in 1996-7 and the budget for 1999-2000 is £100k. At a conference in 1997, ACPO encouraged forces to systematise the payments being made to informants by enacting a 'units' system linking payments directly to the quality of information provided, the risk taken by the informant and the result. Depending on their budgets different forces would allocate different values to units (*The Guardian*, April 4, 1997, 8).

Clearly, particular dangers arise in those cases where informants admit that revenge is a motive. This might be simply part of a personal vendetta but can shade into a more complex attempt by illegal traders to mobilise the 'regulatory power' of law enforcement agencies. The phenomenon by which those engaged in illegal markets might use law enforcement to curtail the activities of competitors has been long appreciated; thus, police might find themselves drawn (perhaps willingly) into a conflict over turf, typically in a competitive drugs market (Skolnick, 1994, 118-9).

This likelihood becomes even greater in the case of participating informants where the management and ethical issues are at their most acute. The Code emphasises that 'informants have no licence to commit crime' (NCIS, 1999a, 1.11) and that they must play only a minor role in any crime and do nothing that suggests they are acting as *agent provocateur*. However, the need for informants to maintain credibility with associates and the organisational realities of law enforcement means that this is impossible to guarantee in practice (cf. Greer and South, 1998, 40-41). For example, a participating informant may well be tasked to gather information on a specific criminal enterprise that is the primary target for some specialist squad. Their concern will be that the informant does nothing to jeopardise that investigation though there will be less concern at breaches of the Code if the object is disruption rather than prosecution. However, if the informant has other illegal enterprises in train they will be tolerated by investigators and, if these activities keep the informant 'on-side' and in funds, then they will be effectively 'licensed' by police in a way discussed in chapter one (see above Figure 1.1). Similarly, Skolnick noted the 'strong tendency' for burglary detectives in the US to permit informants to commit drugs offences and for narcotics detectives to allow their informants to steal (1994, 125-6).

At some point such 'licenses' may tip over into full-blooded efforts at entrapment. For example, a series of 'investigations' and prosecutions between 1990 and 1994 in the UK provided an object lesson in the dangerous scenario of police handler and participating informant working together to entrap targets. In summary, Graham Titley was recruited as an informant by the Midlands Regional Crime Squad when he was arrested in 1990 and thereafter handled primarily by DC Ledbrook. The cases centred on counterfeiting operations and a crucial element was the payment to Titley of several rewards by American Express. Despite the collapse of the first prosecution in 1993 when the judges criticised the bad faith of the prosecution, the CPS pressed ahead with two further prosecutions. The first was dismissed by the judge while the eight defendants in the second, unaware of the full picture, pleaded guilty. For offences for which they might have expected to receive ten years imprisonment they were given suspended sentences and the judge condemned the police operation as 'skulduggery and an abuse of power':

> This was a scandalous, corrupt incitement which led to the defendants being fitted up to commit crimes none of them would have dreamed of committing otherwise. At least one serving police officer from the Midlands Regional Crime Squad knew or believed that the informer was acting as an agent provocateur. The informer was not only acting as an agent provocateur, but was doing so when he was being handled by a serving police officer... (Rose, 1996, 201-7)

The logs of meetings with informants provide the basis for assessments of their effectiveness that contribute to attempts to weed out the unproductive. The figures in Table 8.3 are provided by the police themselves and therefore need to be taken with a pinch of salt, for example, the values put on drugs seizures are notoriously arbitrary and there has been no independent evaluation of the contribution of paid informants compared with other sources of information in leading to these results. While it can also be argued that the payments made to informants reflect only a fraction of the real costs to police of using them, the figures may well underestimate the contribution of informants since these figures record only official payments to registered informants. The pressure on police to make informal payments can be aggravated by frequent delays in obtaining official authorisations (Norris & Dunnighan, 2000, 398).

Table 8.3   Payments to informants

Merseyside Police: payments to informants and results - monthly averages:

	April '94 - Feb. '95*	June - Oct '96**
Total Paid	£5,240	£6,600
No. of arrests	43	66
Value of stolen property recovered	£76,190	£62,000
Value of drugs seized	£71,430	£172,000
No. of recovered firearms	1.2	3.4

Source: adapted from *Cooper & Murphy, 1997, 3; **Barton & Evans, 1999, 41.

Undercover operations

The deployment of undercover means has developed unevenly in North America and Europe. The Anglo-Saxon model of community-based policing reacting to citizen complaints that dominated the nineteenth century was modified in the US by more extensive use of proactive and undercover methods in the twentieth. This started in the area of political policing both at federal and local levels (see chapter five) but is now concerned primarily with 'organised crime'. Gary Marx explains the shift by reference to changing crime patterns and priorities, the lobbying of moral entrepreneurs for new measures that legitimated the tactic, for example, against drugs or corruption and legal developments. Central were the judges' efforts to safeguard suspects' rights against coercion that led, in turn, to increased use of deception and technical developments that facilitated audio and visual surveillance (1988, 36-59; Fijnaut and Marx, 1995, 14).

In Europe it has been conventional wisdom to distinguish the Anglo-Saxon model that developed in Britain from 'Continental' systems. While the latter were based on Napoleonic state-centred systems deploying undercover means from the outset, the former explicitly rejected this model in favour of policing by consent by 'citizens-in-uniform' (e.g. Armstrong and Hobbes, 1995, 176-8). Although the distinction can be exaggerated (Fijnaut and Marx, 1995, 7-10) there were other reasons why, certainly for the period 1945-90, undercover means were less deployed in Western Europe, for example, the rejection of methods employed by the Gestapo. Regarding the UK and white collar crimes, specifically, Levi points out that undercover operations have been used more in the private sector than in the public where their limited use can be explained, first, by the conservative police culture, including opposition to 'foreign' ideas, second, by the fact that deploying resources on reactive investigations was easier to justify and, third, the legal uncertainty as to the admissibility of evidence (1995, 195-211). The relatively short life of the Metropolitan Police's postwar 'Ghost Squad' (Gosling, 1959, 204-5) serves to illustrate Levi's argument.

During the 1990s, however, the gap between Europe and North America has been reduced, in part because European resistance to the proselytising of the DEA has weakened in the new post-Cold war world (Nadelmann, 1993). The process of legitimising undercover operations in the UK (other than in political cases) began, suggested Armstrong and Hobbes, in 1985 when they were directed specifically against football hooligans. The new tactics of infiltration would have been expected to meet with greater criticism had they not been targeted at a sub-group of the population so lacking in credibility amidst the moral panic of soccer-related disorder (1995, 190-91). Similarly, undercover operations were used by Director Clarence Kelley to re-legitimise FBI operations after the traumas of the mid-1970s congressional investigations of the Bureau's illegal countering and disruption of political groups. Between 1975 and 1979 there was a significant decline in the FBI investigative caseload (mainly of domestic political intelligence cases) and an attempt to stress quality, for example, re-orienting the FBI's effort towards, first, organised crime, and then other white-collar crime targets that Hoover had systematically avoided. The number of FBI undercover operations increased from 53 in fiscal year 1977 to 316 in fiscal year 1983 (Poveda, 1990, 151-2, 162fn14). While it is clear that undercover techniques are now used more also in the UK, reliable figures on their precise extent are not routinely available, for example, the annual reports of the National Crime Squad refer to undercover techniques but do not give numbers. It was reported that, during 1994-5 there were 384 Metropolitan Police operations involving undercover officers, resulting in 474 arrests (*Metropolitan Journal*, 1996, 9).

It is important to distinguish informants from undercover operations although both offer similar promise and problems to police - informants are relatively cheaper but in certain circumstances the use of undercover officers will be contemplated. For example, it may prove impossible to recruit informants within a specific criminal group and, if their technical counter-surveillance is sophisticated, then inserting an undercover officer may be the only way of obtaining information. In New York most agencies clearly prefer using undercover officers and therefore the former will be used to 'introduce' the latter. This may be easier at street level while with more sophisticated targets it will be harder to introduce undercover officers and so informants remain essential. Other reasons for 'washing-out' informants is so that they do not have to appear in court or be placed in danger.

The 1999 ACPO code defines an undercover officer as:

> a specially trained law enforcement officer working under direction in an authorised investigation in which the officer's identity is concealed from third parties by the use of an alias and false identity so as to enable: infiltration of an existing criminal conspiracy; the arrest of a suspected criminal or criminals; the countering of a threat to national security, or a significant threat to community safety or the public interest. (NCIS, 1999c, 1.12.1)

What distinguishes this from the use of participating informants, therefore, is that the investigation is carried out by an employee of the law enforcement agency. There is an apparently limitless variety of the kinds of undercover operations in which law enforcement might indulge in order to identify offenders, and the limits on those that are actually used will be set by a variety of factors including the permissiveness of the legal culture, the extent of resources and the freedom for police to indulge their inventiveness. Gary Marx, in his ground-breaking US study (1988), discusses a range of types and dimensions; it is not possible to consider them all here but some are particularly important, for example, the extent of prior intelligence and the specificity of target selection. Often, undercover work involves cases in which intelligence leads to the identity of some person or group as planning some crime or involved in some continuing illegal enterprise. If possible, an undercover agent is introduced to the enterprise in order to gather evidence or to prevent or disrupt the crime. However, there are other types, there may be information about a pattern of crimes but no information on perpetrators in which case police may act as decoys seeking 'victimisation' in order to identify the perpetrator. More controversially, operations might be conducted less because of but rather, in search of intelligence. If this is targeted against a specific person then it may amount to an abuse of authority and harassment if it is based on a simple desire to find *some* evidence of criminality against someone who is believed to be 'ripe'. Finally, there is integrity-testing in which there may be neither specific intelligence nor targets but opportunities to commit crimes will be offered widely just to see who takes them up. Such tactics are common in the private sector where the aim is also to deter by encouraging a 'myth of surveillance' (1988, 68-71).

Different kinds or undercover work have different objectives. Marx identifies three main goals: intelligence, prevention and facilitation, though in complex operations these may co-exist, be sequentially linked and even conflict (1988, 61-67). The first may occur either before or after any specific criminal event. Deploying undercover officers is not widespread in law enforcement prior to a crime because there is greater pressure than in, say, security intelligence, to develop specific cases but officers may be infiltrated into specific locations or organisations where it is believed criminal activity occurs. The objective will be to develop intelligence in order to target specific suspects against whom evidence may be gathered. If, after a crime has been committed, no information is available or technical surveillance is impracticable, then the essentially passive use of an undercover agent as an information-gatherer is relatively unproblematic. But 'human' agents differ significantly from technical means because they may become highly active and the ethical dangers increase because they necessarily form personal relationships with suspects and third parties. For example, Colin Stagg was suspected of killing Rachel Nickell on Wimbledon Common in front of her young son and an undercover policewoman befriended Stagg and pretended not just to enjoy listening to his sexual fantasies but threatened to end their 'relationship' if he did not admit the killing. At Stagg's trial these conversations were treated as 'interviews' and excluded by the judge under s.78 of PACE because they amounted to 'a skilful and sustained enterprise to manipulate the accused' (Maguire and John, 1996, 327).

One of the original goals of 'new police' was the prevention of crime and part of the logic of uniformed patrolling was that it prevented crime. Preventive operations by undercover means have been used more extensively in the past by political police seeking to disrupt the activities of groups targeted as 'subversive' or 'terrorist' (Gill, 1994, 191-200). Now, they are seen as more attractive by law enforcement, for example, they will almost certainly be cheaper than gathering evidence followed by the arrest and prosecution of perpetrators. Also, there is less chance of the undercover role being exposed, risking either the safety of the agent or the integrity of any subsequent prosecution because of allegations that she was an *agent provocateur*. On the other hand, in these circumstances there is no testing of the legality and propriety of police actions and the criminal enterprises concerned may simply re-group and cause even greater damage later elsewhere. The third possible objective is 'covert facilitation' which, in contrast to the preventive logic of policing, is used to apparently *expand* criminal opportunities. This may take different forms depending on the role played by the undercover agent: s/he may be 'victim' - for example the 'decoy' drunk with fat wallet - or co-conspirator where the agent poses as a 'fence', 'hitman', drug dealer or pornographic bookseller.

The debate as to the propriety or otherwise of these tactics goes to the heart of the nature of democratic policing. Anglo-American systems have traditionally relied on citizen-initiation that has the advantage of vesting *some* control in the public over what the police do and police organisations are geared primarily to reactive rather than proactive policing; indeed, it has been argued that such 'minimal policing' would ensure the democratic form (Kinsey *et al*, 1986, 186-215; Johnston, 2000, 35-50). Yet the consequences of such a policy are highly discriminatory:

> Under a libertarian commitment to reactive enforcement as the democratic ideal, pandemic class inequality in criminal justice is inevitable. Because upper class crimes are largely invisible - as crime in the suites they largely occur in private space, while working class or street crimes disproportionately occur in public spaces - proactiveness becomes a necessary, though not sufficient, condition for equality under the law. (Braithwaite *et al*, 1987, 16)

This argument is illustrated by examples from a number of areas - product safety, sexual harassment, illegal waste disposal, environmental protection, occupational health and safety, consumer protection, and employment discrimination. In one specific area, the bribery and corruption of public officials, Braithwaite and his colleagues argue that enforcement policy should be explicitly reformulated so as to require covert facilitation (1987, 16-26). As a policy recommendation this was criticised both because it is extremely unlikely that such a reversal of normal law enforcement priorities could occur and, even if it did, that the rights of citizens would be safeguarded (Baumgartner, 1987, 61-9) but to the extent that law enforcement attention *is* shifted towards 'crimes of the powerful' then undercover policing will become both more significant and controversial, if only because those it ensnares will have greater resources with which to challenge their capture.

The fact that undercover operations are, as a matter of fact, being more generally used obviously does not resolve the outstanding issues of the ethics involved and whether this greater use should be challenged and, if so, on what grounds. Undercover tactics necessarily involve deception and it is possible to argue that states should not use such tactics since they reduce state agents to the same level as the criminals they claim to be countering. However, it would have to be acknowledged that the consequences of such a policy would be highly unequal, for example advantaging the powerful over the powerless (cf. Braithwaite *et al*, 1987) and might be highly damaging. Therefore, discussion has centred more on the conditions under which undercover tactics are a legitimate tactic and the controls that are necessary to minimise the harm that might follow from their use.

Gary Marx has suggested a number of issues that should be considered in deciding whether or not undercover tactics are justified: that the crimes concerned would be seriously harmful, that non-deceptive means are unavailable, that the option of undercover means has been subject to some democratic decision and publicly announced, that the strategy is consistent with the spirit as well as letter of the law, that the eventual goal is to invoke the criminal justice process so that the deception can be made public, that it is proposed for crimes that are clearly defined, that it can be reasonably concluded that the crimes that result are not artifacts of the tactic, that there are reasonable grounds for concluding that targets are involved in the commission of equivalent offences regardless of the undercover operation and that there are reasonable grounds for concluding that a serious crime will be prevented (1988, 105). Braithwaite *et al* offer an even narrower set of conditions that should be applied to the specific use of covert facilitation (1987, 35-7; cf. also Bok, 1986, 265-80).

In considering the UK case, it is instructive to consider how satisfactorily the Code of Practice on undercover operations published by the police and customs in September 1999 deals with these issues. The Code certainly addresses some of Gary Marx's list: undercover tactics will be approved if other means are impracticable or would be unlikely to achieve the end (NCIS, 1999c, 2.2), the 'primary purpose' is to bring offenders before the courts (1.6), and there may be no incitement of offences that would not otherwise have been committed nor attempts to entrap people who were not otherwise disposed to commit such offences (1.10).

However, even as a Code, the guidelines fall far short of what is desirable in other important respects. The main problem is the lack of clarity in the definitions of what might be targeted and great ambiguity about the criterion of 'seriousness'. For example,

> Authorisations for undercover operations will only be given in connection with *national security*, for the prevention or detection of *crime*, for the maintenance of *public order*, for the maintenance of *community safety*, in the case of a significant *public interest*, or in co-operation with foreign law enforcement agencies in these matters. (1.5, emphasis added)

Elsewhere in the codes some of these terms, for example, community safety are defined very broadly (see p.216 above). But others are not included, for example, national security, that is also defined very broadly in the UK (Gill, 1994, 98-100). Taken together, the overall impact of the generality of these definitions is that it excludes virtually nothing that police might be called upon to handle, including many activities that may not be breaches of the law at all. In other places, the Code does suggest, in contradiction, that seriousness is a criterion, for example,

> Undercover operations will only be used by the law enforcement agencies where they judge such use to be proportionate to the seriousness of the crime being investigated…(1.7)

and,

> …the authorising officer must be satisfied that the deployment of undercover officers is likely to be of value in connection with national security, in the prevention or detection or *serious* crime…(2.2, emphasis added)

Such ambiguity in an official document is profoundly unsatisfactory but it does betray the primary motivation in issuing this and the other codes. The most detailed sections of the codes relate to the authorisation procedures including required ranks and review periods - there is nothing wrong in this but the *specificity* of the procedural aspects of the codes contrasts greatly with the *generality* of the substantive aspects. The objective of this is to grant law enforcement agencies the widest latitude for their operational choices while seeking to limit their accountability in courts and elsewhere to relatively narrow procedural issues.

This reflects traditional state strategy in the UK, that is, to maximise the discretion of state officials but it is especially dangerous when applied to an area where the potential for the corruption of state action and damage to third parties is considerable. However appropriate and legal the intended outcomes of undercover action are, though they can be very difficult to measure (Marx, 1988, 108-28 discusses in detail), the technique is especially vulnerable to producing unintended consequences. These may damage specific targets, third parties or the undercover agents themselves. If targeting is specific and the undercover operation produces evidence that will support a prosecution, then fine, but if it is less specific and fails to produce substantive information from which evidence can be developed then dangers emerge. Agents' desire to get at least some 'result' may lead them to seek to damage the target's businesses whether or not they are actually illegal ('he had it coming...'), or the fact of an investigation may be leaked to the press as a form of disruption in such a way that reputations are damaged even if no formal action is taken.

The UK Code acknowledges the dangers posed by undercover operations to third parties and says that practicable measure should be taken to avoid collateral intrusion into their privacy (NCIS, 1999c, 1.8) but this is barely credible in the actual context of many operations. Gary Marx illustrates the specific financial, psychological, physical or reputational damage that may be caused to third parties apart from the more general betrayals of privacy and trust (1988, 142-52).

Agents themselves are also vulnerable to both causing and suffering damage. It is acknowledged that policework is distinctive for the fact that discretion is greatest at the lowest levels of the organisation, for example, because of its low-visibility and the permissive nature of the police mandate. To the extent that this is true for uniformed patrol officers and even truer for plain clothes detectives then its apotheosis is undercover work. Not only is the undercover agent freed from the constraints of superiors, badge, uniform and shift work but they must actually work at conveying the opposite - that they are reliable confidants of those involved in illegal activities. As such they have to decide how far they are prepared to go in order to maintain their credibility rather than risk the operation and, possibly, their personal safety. This situation, in long-term operations rather than in short-term 'buy-busts', gives rise to a variety of social and psychological risks for agents, for example, developing genuine affection for targets, indulging in serious substance abuse, coming to doubt the legitimacy of the operation, living a lifestyle that is difficult to abandon (or that would be impossible for a police officer) (Marx, 1988, 159-79).

Even if such damage to agents is avoided, the damage to the criminal justice process may be extensive if agents behave improperly and evidence is fabricated - various miscarriages might occur: if fabrications are revealed and prosecutions collapse then the 'factually guilty' may be freed; if they are not, the innocent may be convicted. Once significant resources are committed to an undercover operation those involved may develop a fierce determination to get a 'result'. For example, several trials of those alleged to have been planning violence at soccer matches in the UK collapsed amidst revelations that police log books had been tampered with and their statements had been falsified (Armstrong and Hobbes, 1995, 181-3).

What controls can be exercised to try to minimise this damage to all concerned? Broadly, these can be examined as internal and external. Internally, clearly there is a need for some set of rules to which agents are expected to adhere though, as with all guidelines, they may actually play a greater role in presenting an acceptable image to outsiders than actually influencing how operations are carried out (cf. Marx, 1988, 181-8; also Braithwaite *et al*, 1987, 117-9). In the FBI's minor contribution to the 1980s Iran-Contra scandal - 'Iranscam' - the FBI agents acting undercover were later found to be not even aware of the Attorney General's guidelines (Poveda, 1990, 175-6). Care must be taken with the recruitment, training and supervision of undercover officers - in the Metropolitan Police, the process can take up to ten months and includes a series of detailed questionnaires, interviews and psychological tests - 40-60% of applicants fail these. The final stage is the National Undercover Training Assessment Course that is run by the Met for all forces. The Met also employs an 'uncle' system so that an experienced undercover officer is on 24-hour call for colleagues who may have problems (*Metropolitan Journal*, 1996, 9-10).

The possibility of some systematic external control of undercover work (rather than the rather random 'controls' that occur when an operation is exposed or a case collapses in scandal) depends on the nature of the legal context. If, as in the UK and US, this is in the form of a code of practice or guidelines, then the possibility of external control is weak, except through judicial rulings in court as to the admissibility of evidence and the availability or otherwise of the entrapment defence (see further below). It has been argued in both those countries that there should be a statutory framework for undercover operations (Marx, 1998, 195-8; Maguire and

John, 1996, 320-1). Within Europe, there are a few countries where at least some undercover techniques are governed by legislation, for example, France and Netherlands. Germany has the most comprehensive legislation that, since 1991 has offered a statutory basis for police infiltration for the repression of organised crime. In apparent contrast to the UK Code, the law restricts the application of the technique to severe offences involving drugs, trafficking in arms and money, offences against state security, professionally committed crimes or crimes committed by criminal organisations (Joubert, 1994, 33). The lack of precision in these categories does raise the issue of whether there is, in practice, much difference here from the UK Code. However, the existence of legislation does, at least, provide more space for political debate and legal challenge than a code of practice.

Another issue concerns the applicability of prior authorisation for undercover operations. In the UK Code authorisation for longer term operations regarding serious crime is required from an assistant chief constable (NCIS, 1999c, 2.6), and for shorter term 'test purchases' and decoy operations from a superintendent (3.6). In other words, no external authorisation is required. It is curious, given that the invasion of privacy represented by an undercover officer is much greater than that of an electronic bug or telephone intercept, that most jurisdictions now have some form of external prior authorisation for the latter but not for the former. It can be argued that this anomaly should be corrected but it would be extremely difficult to construct a workable system. The nature of undercover work, compared with, say, a telephone intercept is richly complex and it would be impossible to specify just what was being authorised except in the most general of terms. There is also the problem of involving judges in authorising the intimate details of policework that their colleagues later may be called upon to judge in a trial; there is a general Anglo-American reluctance to see judges assuming more of the inquisitorial role of judges in Europe (Marx, 1988, 193-5). Therefore, arguing for prior judicial authorisation is probably not very productive given the problems involved and, even if it could be made workable, it would not be desirable if it worked in such a way as to limit the ability of defendants to challenge police behaviour.

Indeed, the Regulation of Investigatory Practices Bill does not envisage prior authorisation by an external official or judge for undercover officers any more than it does for informants, as we saw above. Although the provision of a statutory framework for undercover policework is to be welcomed, the current Bill can be criticised in similar terms regarding undercover policing as it was above with respect to informants. There is one potentially significant difference, however, concerning the mandate of the Tribunal. This is empowered to hear challenges to the actions of the agencies under the Human Rights Act but whereas the behaviour of informants will be excluded because they are not employees, that of undercover agents will be subject to challenge (s.56[6]). Thus a fresh impetus may be given to challenges to state actions that are considered to border on entrapment.

Hitherto, there have been two main strands of judicial practice in evaluating undercover operations in the context of the defence of entrapment. In the US, the courts have adopted a 'subjective' test in which the crucial evidence is seen as the 'predispositions' of the accused to indulge in the kind of activity into which they were attracted by the undercover operation. Thus the entrapment defence will only be allowed where the disposition to commit the offence has been implanted by a government agent. This contrasts with the 'objective' approach - as in Australia and Canada - that asks whether an ordinary person in the position of the accused might have been persuaded to commit the offence. Here, regardless of the disposition of the accused, the question of the state's conduct is more central - has it acted in such a way as is likely to create an offence? (Robertson, 1994, 805-16; Kleinig, 1996, 153-6).

In the UK the situation has fluctuated over the past thirty years: in *Ameer* (1977) the case collapsed because the judge was satisfied that a police informer had played a crucial role in drawing the defendants into a criminal enterprise they would not otherwise have contemplated. As a consequence, charges were dropped against 26 other people awaiting trial on the evidence of the same informant and police officers and a corruption inquiry was instituted. However, the impact of this decision was overturned by *Sang* (1979) in which it was decreed that the trial judge

> has no discretion to refuse to admit relevant admissible evidence on the ground that it was obtained by improper and unfair means. The court is not concerned with how it was obtained... (quoted in Robertson, 1994, 809)

The Police and Criminal Evidence Act 1984 (PACE) reversed this to the extent that trial judges were granted discretion to exclude evidence, the admission of which 'would have such an adverse effect on the fairness of the proceedings that the courts ought not to admit it' (s.78[1]). The fact that this *may* be applied has predictably led to varying decisions, Sharpe argues that by applying the pre-PACE common law to this section the judges have essentially continued to validate the police use of deception (1994, 793-804), but others have argued that the *Smurthwaite* (1993) decision, by reiterating some of the tests outlined in *Ameer*, provides a clearer basis for the use of PACE s.78 in controlling entrapment (e.g. Robertson, 1994, 809-11). Being typically pragmatic, the approach of UK judges cannot easily be fitted into either the 'subjective' or 'objective' positions referred to above. It leans towards the former but is concerned also with what type and how much pressure police apply.

Corruption within law enforcement is, ironically perhaps, a particularly appropriate arena for the use of undercover techniques. The special powers available to state agents and the significant rewards and punishments that are available between enforcers and those they seek to regulate provide fertile ground for corruption.   The problem is aggravated by the fact that those concerned have particular skills and privileged access to information that enable them to protect their own illegal activities and detect efforts to surveille them.   Therefore undercover techniques will be required. This has long been recognised in the US, where, as we saw in the case of New York (see chapter five) there has been a regular cycle of corruption scandals and inquiries that has at least led to the realisation that corruption can be systematic. In the UK, by comparison, the official line was for many years that corruption was rarely a problem in policing and, if it occurred, it was because of a 'rotten apple', therefore the problem was dealt with by the removal of the corrupt officer(s) if and when allegations were upheld. Although major corruption scandals were exposed in the Metropolitan Police from the late 1960s onwards (Cox *et al*, 1977) this position has only recently been modified, though not abandoned, as the tide of allegations has grown (e.g. *Guardian*, Dec 14, 1998, G2, 2-3). The Police Inspectorate conducted a review (HMIC, 1999b) and a simultaneous ACPO study led to the initiation of a number of measures, including the offering of witness protection safeguards to police who informed on dishonest colleagues and random integrity testing (*The Independent*, July 16, 1999, 11).

By this time the Met had already introduced its own anti-corruption branch with 200 officers and 80 ongoing investigations. Its head reportedly accused senior detectives of sabotaging undercover operations in return for payments of up to £50k, with criminal entrepreneurs employing former police officers as go-betweens in order to protect their enterprises

(*Intelligence*, N.95, March, 1999, 13).   Merseyside Police suffered embarrassment in 1999 when the last of a series of 'fly-on-the-wall' TV documentaries showed the arrest for corruption of one of the featured detectives. It emerged that Merseyside had actually set up its own specialist Professional Standards Unit five years earlier. Once a specialist squad is established for the investigation of any kind of crime the first consequence will be an apparent increase of crimes of that kind since, now, there are officers with a specific bureaucratic interest in investigating it. So, the advent of these squads confuses the evaluation of whether there has been some objective increase in police corruption or, especially in the UK, whether previous official complacency simply masked its extent.

However, it is possible to identify certain new factors in the 1990s that potentially changed the nature of corruption, if not necessarily increasing it. Whereas the London cases in the 1970s were partly about police officers enriching themselves from the proceeds of illegal markets in products such as pornography, what was at the heart of the most notorious miscarriages of the 1970s and 1980s was not the personal enrichment of the police but their use of violence and fabrication of evidence to secure the conviction of those accused of serious crimes, most notably the succession of Irish cases. In the 1990s, as policing has shifted more attention towards the policing of illegal markets, the sheer amounts of money available in those markets has massively increased the temptation for officers, again, in those areas that can be formally defined as 'victimless' (Newburn, 1999, 16-25 provides a more systematic review of the causes of police corruption).

However inevitable the deployment of undercover methods might be, if a serious attempt to detect the corruption of law enforcement is to be made, it must be remembered that the unintended consequences of their use apply even more.   Policing does not take place in an unambiguous moral and legal context and a targeted investigation of almost any police officer is likely to produce evidence of the infraction of *some* rule. Therefore the technique can be a dangerous weapon in the context of intra-agency feuds or might even be used as a means of dealing with problems that should have been handled as personnel issues. There are costs also for those employed in anti-corruption units - the psychological and social costs of carrying out undercover work may be magnified when the targets are one's own colleagues. However much some police may be determined to expose the wrongdoers in their midst, being a member of a 'rat' squad has never been the route to comfortable relations with colleagues (cf. Marx, 1995). In the most pessimistic scenarios, those deploying undercover tactics against officers believed to be corrupt may themselves employ measures of such dubious legality that still further miscarriages of justice may be perpetrated (precisely such allegations are aired in *The Guardian*, March 4, 2000, 1, 8-9). Alternatively, the establishment of specialist anti-corruption squads may be a measure aimed entirely at public reassurance and its members may be unwilling or unable to actually make a significant impact.

Conclusion

The new Codes of Practice issued regarding the police use of informants and undercover techniques are to be welcomed to the extent that they are a genuine effort to regulate policing and open up this highly controversial area for public debate. It would be a major problem if their primary purpose is simply to protect existing police practices from legal or other challenge within the context of the Human Rights Act. Guidelines are all very well but, as publicly accepted by the recent report of HMIC in UK, they may be circumvented (1999b, 4). They are not legally enforceable, and unless they become embedded within a much more rights-oriented police working culture (e.g. Cooper & Murphy, 1997), may be doomed to become simply the latest in a long succession of unsuccessful attempts to control this most sensitive and problematic area of contemporary policing. There has recently been official acknowledgment that current training for handlers and controllers is often 'woefully inadequate' (HMIC, 1999b, 4) and a much more systematic training regime must be instituted with a view to protecting both sources and police from exploitation (Barton & Evans, 1999, 45). But even improved training will not remove entirely the conflicts that can erupt both within and between agencies because of the clashes of personal and organisational interests inherent in the use of informants (Norris & Dunnighan, 2000). All of these changes would be better supported by a statutory framework but the Regulation of Investigatory Practices Bill is not, on the face of it, going to be adequate. The review mechanisms established are too limited to provide a proper audit of the complex dynamics of using 'human sources'.

The new Covert Investigations Commissioner will be responsible for the review of the actions of those giving authorisations for the use of informants, not of the authorisations themselves, and will assist the Tribunal (s.53). The Tribunal may hold proceedings to hear any challenges under the Human Rights Act to the conduct of law enforcement or security agencies and to consider complaints, but the Bill seeks to ensure that it is only the conduct of the officers that is challengeable, not those of the informants (s.56[6]). It remains to be seen whether these procedures can provide any real check on the abuses of the informant system. If the Covert Investigations Commissioner is asked to investigate a particular case by the Tribunal, the premium will be on the quality of paperwork - the audit trail of authorisation. If this is complete then the 'reasonableness' of the police in authorising the use of the informant will be very hard to challenge. The emphasis throughout the new procedures is on management and efficiency, not ethics. For example, before the authorisation of the use of an informant can be renewed after twelve months, there must be a review of their use and what they have produced but there is no reference to ethical issues (s.41).

The only alternative to the difficult task of controlling law enforcement in this area, in the long run, is to reduce law enforcement's currently increasing reliance on undercover techniques; an issue that has already been considered by Gary Marx (1988). Of the strategies he considers, the one that would most likely have a significant impact would be in re-defining certain crimes, for example, the decriminalisation of (some) drugs could have a dramatic effect on the workings of illegal marketplaces and therefore of the methods by which law enforcement seeks to regulate them. Alternatively, more effective crime prevention measures might be considered, in the case of illegal markets, again, the object would be to manipulate the conditions within which 'trading' takes place in order to reduce its extent or the damage it caused. To the extent that, as we have seen, undercover methods have grown in significance in response to the curtailment of some older 'overt' means of investigation, then by removing these restrictions, the process might be reversed. This might be a stronger argument in the US where the impact of judicial decisions has been to curtail significantly the possibility of early arrest for the purposes of questioning whereas in the UK such tactics are embedded within PACE.

In the meantime, there can be no escape from the conclusion that this area of policing will remain significant because of its indispensability for investigating (whether for information or evidence) certain low-visibility activities including the corruption of law enforcement itself. It will also remain highly controversial because the pressures for secrecy regarding police practices run directly counter to the broader requirements for transparency in criminal justice processes.

9 Producing and Using Intelligence

Introduction

All information gathering systems, whether organised formally by the state or an individual person, are prone to 'naive empiricism', believing that more information will automatically lead to becoming better informed. Indeed, precisely the opposite may be true: problems of information-overload are endemic in law enforcement (Tremblay & Rochon, 1991, 278). Evaluation is about deciding what information 'means' and it may, of course 'mean' different things to different agencies working within different contexts: it is a process of 'making sure that you learn more from the information you have obtained than just what other people want you to know' (Wilsnack, 1980, 475). Although those responsible for targeting and gathering will operate with some more or less well-supported assumptions about the activities of specific groups and individuals, 'facts' rarely speak for themselves - whatever is gathered must be evaluated or analysed. Dissemination is the final stage of the intelligence process and is crucial to the reputation of often-small and sometimes marginalised criminal intelligence units (CIUs). Their ability to maintain credibility and thus enhance their organisational project is determined to a large extent by the perceptions of those to whom they disseminate their 'product', whether members of the same or a different agency.

Organising analysis

Analysis has been described as the 'heart' of the law enforcement intelligence process (Peterson, 1999, 1) though 'brain' might be a better metaphor given that it is essentially an intellectual activity. Crime analysis can be defined as the identification of and the provision of insight into the relationship between crime data and other potential relevant data with a view to police and judicial practice. (Interpol, 1998, 9)

As such, it has always been a feature of good investigative work but, as a self-conscious discipline, it has its law enforcement roots in the US in the late 1960s, especially in the work of the President's Commission on Organized Crime (1967). Godfrey and Harris's *Basic Elements of Intelligence* (1971) became a guidebook and its revised edition, published in 1976, showed that a number of federal agencies were using various analytical methods such as telephone call analysis and event flow charting. At the same time a number of public and private agencies started to offer training and by 1980 analytical techniques were starting to be used by larger state and local agencies. This spread was encouraged by federally-funded RISS projects (see also chapter two) making software and training more widely available. A group of US and Canadian analysts established a professional organisation in 1980 - the International Association of Law Enforcement Intelligence Analysts (IALEIA) - and a loosely associated group - the Society of the Certified Criminal Analysts - in 1990 to establish tests and standards for experience, education and training (Peterson, 1995, 3-5). In the UK two separate developments contributed to the spread of analytical techniques, the first was when the Metropolitan Police's SO11 branch bought the rights to teach the courses developed by Anacapa Sciences Inc. and the second, was the development of HOLMES for the management of the large quantities of information generated by major enquiries (e.g. Oldfield, 1988 and see chapter seven).

From the outset, the story of criminal analysis has been bound up with the development of ICT, not always productively. In an ideal world the new policy strategies of intelligence-led policing would have been discussed, refined and implemented *together with* technological innovation that was consistent with those strategies but, in law enforcement as elsewhere, the decisions were taken by different groups of people with different personal and organisational priorities so that any consistency between strategic and ICT innovations was often more by luck than judgement. A clear example was the widespread innovation of computer-aided dispatch systems that, emphasising the central direction and rapid response of patrol officers, conflict with ideas of policing as a more proactive, preventive, or 'intelligent' activity (c.f. Sparrow, 1994, 101-8). Manning similarly points to the difficulty of shifting the strategic centre of policing when it is dominated by reactive policing and the retrospective intelligence that it both arises from and creates (1992, 364-9).

All intelligence training emphasises the fact that analysis is a continuous cycle, in which 'hypothesis testing' (a sub-process of the analysis stage), goes back to the gathering stage and so on round (Kedzior, 1995, 11, fn9; see also Shuy, 1990, 144-6). This is represented in Figure 1.2 by ↑7 (p.22). Such an 'ideal-type' process will of course always be subject to real constraints of time, will and so on. Arthur Hulnick argues that even the notion of 'cycle' endows what is actually a rather messy process with an unreal logic. His preferred model is of a 'matrix of interconnected, mostly autonomous functions' (Hulnick, 1991, 84). Therefore, an important issue for analysts is the extent to which they succeed in modifying prior assumptions; if they adopt a strategy of seeking to 'disconfirm' them, then it is possible that they will provide negative feedback but if they simply accumulate information that supports those prior assumptions and ignore (as 'unreliable' or inconvenient) information that contradicts them, then feedback is positive. By leading to the apparent affirmation of an actually false hypothesis, this can have the most regrettable consequences.

The analytical process has been a 'Cinderella' in the world of intelligence, both foreign and domestic, as analysts have been unable to compete for resources and prestige alongside their more glamorous operational colleagues. In law enforcement this struggle for credibility is aggravated by the youth of the discipline and the fact that whereas investigators are predominantly male and sworn (or, 'badged'), analysts are predominantly female and unsworn (Sparrow, 1994, 112; Dintino & Martens, 1983, 121). Analysing the information gathered is a crucial part of the investigative and intelligence processes but, being pre-eminently an intellectual process involving critical thinking (Peterson, 1999, 4), is one that normally receives lower priority than the 'action' orientation of much law enforcement (e.g. Reiner, 1992). Since 'paper cops' will not face the rigours of the street, it will be easy for their contribution to be denigrated and it is necessary to implement specific organisational incentives if intelligence work is to escape a reputation for harbouring the 'sick, the lame and the lazy' (cf. Dintino and Martens, 1983, 63). The short-term, case-oriented nature of police operations means that, as 'customers' police investigators will be interested much more in current 'tactical' intelligence with some immediate pay-off potential and may be extremely sceptical as to the value of analysing broader or longer-term patterns of activities (and see further below). Worse still, 'analysts' will be seen simply as the inputters of statistics and producers of the monthly reports for headquarters, including colourful charts that state the obvious (Blaney, 1998, 17).

In Canada the analytical sections of the CIUs tend to be organised separately for administrative reasons although the analysts themselves often work as part of investigative or research teams. All of the Units studied contained some mix of police officers and civilians; the practice of incorporating civilians into this area of policework apparently echoing McDonald's judgement that police officers are not best-suited to intelligence work (1981, 706-16). In general in the RCMP sworn officers tend to dominate at the tactical level while the strategic intelligence programme has mainly civilians (Smith, 1997, 19); in 1995 RCMP criminal intelligence in British Columbia had three police and five civilians, CISO had three of each, the Metro Toronto Unit consisted of seven police and four civilian analysts and CLEU had mainly civilian analysts. Those interviewed saw a mix of police and civilians as productive: the latter provide a continuity that police cannot because their career paths require regular movement between jobs. But the counterpoint was the lack of career path for civilian analysts, a problem that the RCMP claimed to be addressing (Fahlman, 1995).

Prior to its merger within the NCS the North West Regional Crime Squad in the UK was based in three main and two smaller offices. The former each had a Development and Research team consisting of four to six police or civilian analysts. Limitations of time and money meant that the then head sought to recruit people who already had relevant training, for example, in the military or by poaching from police forces (Nicholls, 1996). Ten percent of the 300 people working in UK Division at NCIS in 1997 carried out solely analytical work while others might also carry out such tasks as telephone analysis from time to time. NCIS is only recruiting civilians as analysts (Clay, 1997). Dedicated analysts are relatively thin on the ground also in UK Customs compared with the number of staff working in intelligence more generally, for example, four of the sixty in the North West Collection (Parker, 1997).

At the time of this research, strictly speaking North Wales had no analysts although a financial bid was being made for one. One person in the FIB had started work as a clerk in the previous Criminal Intelligence Bureau and her job had changed with the implementation of the new criminal intelligence policy so that she was described by some as the force analyst. But, having undergone just a two week training course with Greater Manchester Police she did not describe herself as an analyst and half of her time was spent on responsibilities as the Data Administrator under the Data

Protection Act. Also in FIB were several civilians who had been recruited as youth trainees and had had no general training in analysis but carried out specific roles such as mapping telephone networks (Hennings, 1997). Merseyside introduced a crime analyst to each of its five areas plus five in the FIB during 1997-98 (HMIC, 1999, para.3.12).

As far as training is concerned, as before, we see a number of parallel, occasionally intersecting, flows in Canada. CISO prided itself on providing

> a cutting edge training curriculum for police officers in the areas of investigation, criminal analysis, mobile and technical surveillance, intelligence gathering and Proceeds of Crime. (CISO, 1995, 3)

As far as 'strategic' intelligence was concerned, finding nothing 'on-the-shelf' in North America, CISO sent students on courses taught in Canada by Don McDowell who was the founder and president of the Australian Institute of Professional Intelligence Officers. He advised on the development of courses and trained the trainers and CISO together with the RCMP, Canada Customs and Edmonton Police Department ran the first homegrown strategic analysis course in May 1995 (Faul, 1995).

CLEU sought guidance from nearer to home, for example, 'buying-in' an intelligence training course for itself and other provincial agencies from Anacapa Science Inc. (CLEU, undated, 2). The RCMP also used Anacapa but then developed its own Criminal Intelligence Analyst training programme. The three main components of this programme are: the theory and practice of the role of analysis within the intelligence process; a specialised topic in which the analyst will subsequently specialise; and acting as witnesses in court. The specialised areas of responsibility are: drugs, economic and computer crime, organised crime, and security offences and criminal extremism/terrorism. The course lasts on average 20-24 months but good students might complete it more quickly. After the first three months training is conducted concurrent with actual assignments.

Only limited training opportunities have been available for analysts in the UK, initially in areas such as financial analysis though the opportunities are greater in larger CIUs. The military have provided some and NCIS developed a national programme during 1997 (Clay, 1997). In Customs the National Intelligence Division sought to develop their own training, initially through bringing in outside consultants such as Don

McDowell. Otherwise, training seems to come primarily in the form of people undergoing instruction in the use of specific software packages, for example, for telephone analysis or, at force level, typically a 3-day course regarding putting information into the force database.

Tactical and strategic analysis

CIUs are generally considered to produce either *tactical* and/or *strategic* intelligence although in practice these can merge. The distinction between them can be traced back at least 25 years (Godfrey & Harris, 1971, 2-3); tactical intelligence is presently defined by the RCMP as:

> principally an investigative tool, (it) is the support given by the analytical unit or criminal intelligence section to operational sections or investigators during the course of an investigation.

While strategic intelligence:

> Largely a management tool, attempts to provide an overview of the scope and dimension of criminal activity, to assist in policy development aimed at providing effective strategies to deal with the overall costs and effects of criminal behaviour on society. (RCMP, 1995, 1)

It might be better to view them as points along a spectrum, for example, Richard Kedzior distinguishes two 'types' of strategic intelligence: enforcement/operational and policy-planning (Kedzior, 1995, 5-6). The latter is clearly identical to the RCMP definition of strategic intelligence while the former relates to:

> the analysis of a crime group, overall criminal activity, or situation which results in the production of a report on that group, activity or situation and includes recommendations for future action. (Kedzior, 1995, 5)

The difference between this and the RCMP definition of tactical reflects no major conceptual difference, rather it is a 'half-way house' reflecting the fact, as we have seen, that most CIU work is concerned with *organised* crime. However, *groups* cannot be prosecuted and therefore as the intelligence process moves nearer to the investigative process it has to be directed more to the activities of prosecutable *individual* cases.

The tactical/strategic distinction, however, is useful in understanding the sometimes tenuous foothold that CIUs have within what are essentially law enforcement bodies. Policing has not been traditionally an intellectual activity:

> The clinical nature of (police) work as defined traditionally and in the police culture as well as its immediacy and situated character, and the structural characteristics of police knowledge, lead to the discreditation of paperwork, files and systematised information. (Manning & Hawkins, 1989, 146)

Since these factors go to the very heart of the analytical process the problem is clear. But there are also important and well-documented distinctions *within* police agencies, best summed up in the distinction between 'street' cops and 'management' cops (Reuss-Ianni & Ianni, 1983). The former will perceive the latter as concerned primarily with the public and presentational face of policing - keeping trouble-makers such as politicians and media at bay - and less with the more immediate investigative and enforcement objectives of the street cops.

The distinction between tactical and strategic intelligence, therefore, is not just a matter of definition or semantics, it has real organisational consequences and permeates the analytical process. Don McDowell has bemoaned the fact that the frequent organisational separation of strategic intelligence activities extends to it becoming intellectually separated from tactical ones with the danger that it culminates in mutual misunderstanding and disdain (McDowell, 1996). In general the tactical will dominate the closer the CIU is to 'the street'. In the RCMP 80-90% of what divisional CIUs produce is said to be tactical whereas most of what is produced at headquarters is strategic (Fahlman, 1995). In Ontario most of the effort of the Metro Toronto Unit goes into developing cases; there has been little effort to produce strategic or 'predictive' intelligence but in 1995 it was working on more general reviews in the areas of Black and Asian organised crime (Sandelli, 1995). Meanwhile CISO aimed to produce just strategic intelligence (Faul, 1995) and in British Columbia about 20% of what the RCMP intelligence unit produced were strategic reviews for senior managers (MacPhee, 1995). CLEU incorporated both emphases: the intelligence produced for the joint forces operations was mainly tactical while the Policy Division produced primarily strategic intelligence (Engstad, 1995).

The numbers and location of analysts in UK law enforcement - usually working in close proximity to investigators - means that their product is primarily tactical, with only limited opportunities for the production of strategic intelligence. At regional level RCIOs would be tasked for information on specific operational targets (Nicholls, 1996) and even at NCIS the 'bulk' of the product was tactical in 1997 although the then recently-appointed head of Analysis was expected to work up the strategic area (Clay, 1997). Nobody in North Wales claimed to be producing strategic intelligence.

The nature of the enforcement environment within which Customs operates compared with the police gives a more clearly-defined space for the production of strategic intelligence. 'Risk' and 'threat' assessments are very much the current language within Customs who, because of their mandate, receive regular, large quantities of information by way of Customs declarations and VAT returns that provide them with a ready-made data base that police do not have. Whereas it was acknowledged that at Collection level almost all work was operationally linked to cases, NID claimed to be providing various forms of strategic advice - policy, legal changes, forecasting trends and environmental scanning - whereas they would produce operational intelligence only as a by-product (Wesley, 1997). Similarly, in the New York DOI, use is made of system analysts, auditors and accountants to conduct risk analysis by flow-charting the operations of an agency to see where decisions are made and where are the crucial weak points in terms of corruption or fraud (Burke, 1999).

The relative weakness of police agencies in developing strategic intelligence has been cited as one of the reasons for involving security intelligence agencies increasingly in law enforcement matters. It is said that they have greater experience with this kind of work and that therefore they can make a vital contribution since police agencies will inevitably face the shorter-term pressures to develop intelligence with a view primarily to developing cases. This argument can be challenged, for example, the Security Service has certainly been in the business of running much longer-term intelligence operations many of which would not culminate in a court appearance, however, it was still operational in the sense of targeting specific individuals and groups. As such, their work is less different from that of the police than has sometimes been argued (e.g. Farson, 1991).

Analysts and investigators

In addition to the differences in organisational status between analysts and investigators discussed above, there are other issues regarding the relationship between these two functions. Several interviewees commented that the most desirable state of affairs was one in which each investigative team had an analyst permanently allocated but that this was not possible. If it were, then this could maximise the utility of the analyst's position, constantly evaluating what the investigators collect in the light of the 'store' of knowledge and then feeding this back to investigators by way of a fresh collection plan which can then act as a way of testing any analytical hypothesis (Kedzior, 1995, 11). Equally, however, there is a delicate balance to be struck: analysts may be too easily sucked into the investigative process, for example, preparing evidential material for court rather than intelligence evaluations (see Peterson, 1995, 6-8 for a more general discussion of the relationship between analysts and investigators).

To the extent that the smaller New York agencies within this study employ specialist analysts, they are also most likely to be working in direct support of specific investigations and consequently producing tactical rather than strategic analyses. For example, one police unit with nine analysts primarily supports investigators in the field - 70:30 typifies the balance between tactical:strategic. Similarly, each Organized Crime Task Force investigative team in New York has one analyst who tends to stay in the office doing the telephone checks that investigators cannot do for themselves. These can include toll analysis and inquiries to other agencies and MAGLOCLEN.

NYPD is different in a number of respects: because it is much bigger it can afford more specific analysts - there are 36 in the Intelligence Division - but they are all sworn police officers. The Division would prefer to recruit civilians but the political imperative to recruit more police means there has been no budget provision for them (Oates, 1999). The Division is also distanced from investigative pressures having been divorced from enforcement duties when it was (re-) established in 1996 with the intention of preventing turf battles with the Organized Crime Division. In the main prosecutors' offices in New York there are only a few analysts. Outside of the sections carrying out longer-term investigations analysts as such can be seen as 'a bit of a luxury' by prosecutors (Heimer, 1999). Prosecutors want 'evidence', not intelligence. The current obsession with performance measurement aggravates the long-standing difference here:

In intelligence, the measurement of success is the impact on the problem *regardless* of arrest, prosecution, or incarceration statistics. In investigations, these very statistics measure effectiveness and efficiency. (Dintino and Martens, 1983, 81, emphasis in original)

Analytical methods

There are now a large number of analytical methods available for criminal analysis: Peterson (1995) describes 26, though the number routinely in use is far fewer. The National Intelligence Model promulgated by NCIS in 2000 refers to nine (this Model is discussed in chapter ten). Read and Oldfield (1995, 45-6) identified eight forms of analysis: three of them described as strategic: crime pattern (CPA), general profile and crime control methods. The first of these is now commonly in use among British police forces, though with mixed results. At one level Geographical Information Systems (GIS) fed with crime data simply provide a high-tech (and very expensive) way of doing what police have long done - put pins representing crime occurrences in maps and observing the resulting clusters. Contemporary CPA systems are capable of far more than this, for example, as Compstat does in New York (Silverman, 1999, 103-05), but their actual utility may be modified by high costs, lack of 'fit' with the available data and insufficient analytical skill in deploying the programme (Read and Oldfield, 1995, 27-32). Crime control methods analysis seeks to evaluate the impact of previously-applied techniques and tactics.

General profile analysis aims to identify the typical charac-teristics of those perpetrating certain crimes; for example, Customs make use of such profiling in terms of suspected drugs couriers at ports along with other risk analyses of their originating countries. There is no specific method for this and it may be little more than glorified stereotyping; indeed it may degenerate into meaninglessness: through examining court cases in the US in which officers used traits from their agency's profiles to justify making stops, David Cole compiled a list, of which the following are examples:

Arrived late at night...Arrived early in the morning...Arrived in the afternoon
One of the first to deplane...One of the last to deplane...Deplaned in the middle
No luggage...Brand-new luggage

Carried a small bag...Carried a medium-sized bag
Carried two bulky garment bags...Carried two heavy suitcases...Carried four pieces of luggage
Made eye contact with officer...Avoided making eye contact with officer
Suspect was Hispanic...Suspect was black female (1999, *Harpers' Magazine*, October, 1999, 26)

Read and Oldfield (1995) also identify five forms of operational (tactical) analysis, three of which relate specifically to the investigation of specific cases and are essentially retrospective. This reflects the continuing predominance of reactive policing and also, more specifically, that the investigations of a small minority of offences can become very high profile due to the nature of the victim or the offence. Case analysis involves charting chronologically the activities of all relevant individuals and events surrounding a serious offence. This is used then to identify gaps in knowledge about the offence to inform subsequent investigation (see Figure 9.1). Investigations analysis is related in that it involves the evaluation of the investigation itself with a view to identifying incomplete areas. Third, specific profile analysis involves constructing a hypothetical picture of the offender on the basis of the information gathered from the scene of crime, witness statements etc. Commonly known as offender profiling, this has become a highly controversial area, for example, its use in the Stagg case, and claims for its use have probably far outstretched its actual value - in part because of its fictional representation in the TV series *Cracker*.

Offender profiling also provides for one variant of the fourth form of analysis - comparative case analysis. This seeks to detect similarities between offences that point to the same perpetrator(s). Again, this is not new in the sense that even before the days of computers, police maintained paper indexes of *modus operandi* but this is clearly an area where the ability of computers to store and process quickly very large quantities of data potentially makes a big difference. This is an area of policing where strategic innovation has matched the development of ICT capacity: the advent of software for the comparison of crimes in terms of location, time, *m.o.* and so on lends itself well to the shift in strategy since 1993 of targeting, first, offenders and, second, crime 'hot-spots'. Yet, there are important reservations, for example, the way of recording *m.o.* in high

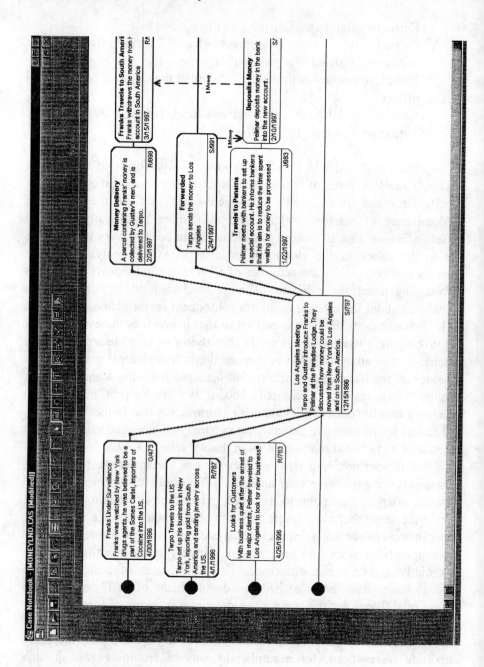

Figure 9.1 Case Analysis
Source: i2 Case Notebook

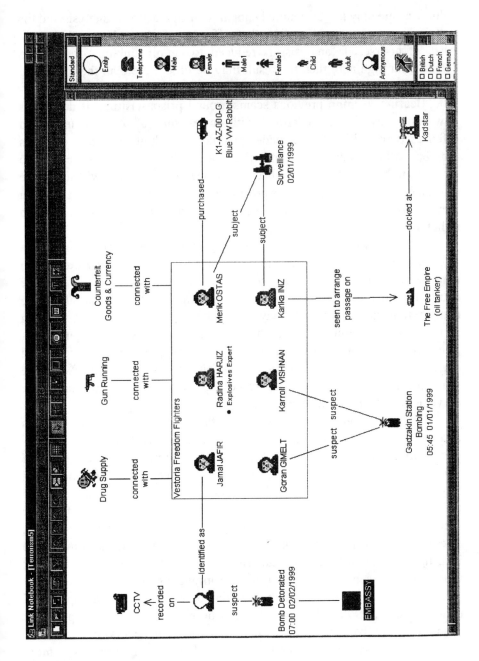

Figure 9.2 Link Analysis
Source: i2 Link Notebook

volume crime may not be detailed enough to provide proper analysis and the system rests on the assumption that offenders do not change their *m.o.* Also, of course, if 'analysis' is reduced to simply linking offenders to offences through *m.o.* then the result may be little more than the re-cycling of known offenders through the system (cf. Read and Oldfield, 1995, 13-15). Bearing in mind previous scams around prison-visiting, if detectives now visit those in prison armed with correlations of *m.o* and uncleared offences their ability to 'persuade' prisoners to admit to further offences may be further enhanced. If and when those admissions are fed back into the system then what may have been an erroneous conclusion will simply be reinforced. This method also has prospective potential when it is used with a view to prevention, for example, increasing patrols or CCTV surveillance in a 'hot-spot'.

The aim of Offender Group Analysis is to collate all the available information on people believed to be involved in criminal activities and to map these in such a way that hypotheses can be generated regarding the nature of the relationships between people and activities. As such, like comparative case analysis it may be used in the analysis of an ongoing series of offences but it has also has potential for prospective intelligence. The associations or, 'links' found between people, locations, vehicles and events can be presented in matrices or on charts (e.g. see Figure 9.2). The same techniques are used with respect to tracking in flow charts the movements of commodities and series of events.

The growing sophistication of the software available for this relational analysis has, again, led to some over-estimation of its possibilities. There is no doubt that the programmes can be of immense value in manipulating very large quantities of diverse data and the visual representation of complex social networks. This procedure has a sound theoretical basis in anthropology, for example, for the depiction of the structures of relations that form the basis for social action (Scott, 1991, 2-5). These can then be used for developing hypotheses to inform future investigation and information-gathering and also for the identification of those who are central characters or who might otherwise be strategically located, for example, providing the sole link between different groups (Ianni & Reuss-Ianni, 1990, 67-75). This can be significant in developing, for example, disruptive techniques against a network (e.g. Klerks, 1999, 4).

As we saw in chapter three, the network form provides a more accurate conceptual basis for the analysis of crime and to the extent that these programmes aid in the mapping of criminal networks, they represent a great advance over the days, especially in the US, when analysts focused on locating individuals within what were assumed to be static, often hierarchical, structures (Ianni & Reuss-Ianni, 1990, 77). However, important caveats must be lodged. Charting or mapping a network is an essentially *descriptive* exercise that can provide a sophisticated representation of much quantitative data as to the existence and /or extent of relationships at a particular time between people, events and so on. Even here, care must be taken as to what is being counted since information collection (especially when it is covertly targeted on the 'usual suspects') will be partial and biased (Klerks, 1999, 6). However, analysing the network requires a further and separate process. This is directly analogous to the methodological maxim: 'correlation is not causation'. *Quantifying* the relation between two suspects tells us something but it does not of itself tell us anything about the nature or *quality* of that relationship (Ianni & Reuss-Ianni, 1990, 78).

One of the original contributions of analytical software products such as those produced by i2 was their ability to provide visual representations of raw data and maps of social networks from the links between different data sets relating to names, vehicles, locations, incidents and so on. Software has now become far more sophisticated and plays an integral part in the analytical process, helping to frame the Who? What? When? Where? How? and Why? questions that should be asked during the investigative and intelligence processes. Some of these questions may be answered by innovative linking of existing databases; others will require more extensive research. i2's Analyst's Workstation, for example, was designed to support UK police forces in their response to the Crime and Disorder Act 1998 and, through the application of a variety of analytical techniques, supports a proactive, intelligence-led approach to crime problems and crime reduction strategies. The amount of research involved in obtaining the necessary qualitative information may not be feasible in routine criminal investigations but in major investigations and for strategic purposes the study of the content and context of relationships is required to understand better the power and affinity structures that constitute criminal networks (Klerks, 1999, 8). Mapping the social networks that underpin the operation of illegal markets is certainly a *necessary* precondition to intervening in them but it is not sufficient; just as social scientists must beware slipping too easily into believing that network maps actually *explain* events (Dowding, 1995, 150-8) so law enforcement agencies must beware falling into the same trap. Analysts may well lack the training and/or the time to think about the implications of the relationships they present.

Organising dissemination

This is the final stage of the intelligence process and is especially significant since its effectiveness largely determines the credibility of intelligence work (↑8 in Figure 1.2, p.22). It is clear that CIUs have a somewhat marginal existence within police organisations; they are small and their activities are often viewed with suspicion by operational police (in some cases they are seen as covers for internal disciplinary units). The method by which they can safeguard their position is by producing intelligence that is valued by others within and without the police organisation. Thus it is at this stage of the intelligence system that CIUs attempt to overcome the resistance of both operational and managerial sceptics; units take care in what they disseminate because their own performance will be under surveillance and evaluated largely on the basis of what they produce (cf. Dintino and Marten, 1983, 128-33). For example, a review of the RCMP criminal intelligence programme in 1996 was very critical:

> The CI (criminal intelligence) program is not adequately meeting the needs and expectations of its clients…Clients perceive program products to be of little tactical value. They feel that sharing of information/intelligence is too often incomplete or untimely. In their view, the National Crime Data Bank (NCDB), the automated intelligence system of the CI program, is inefficient, nonproductive and costly. (RCMP 1995/96)

All CIUs will first seek to maintain or enhance the 'organisational memory' (↑9 in Figure 1.2, p.22) and so all intelligence analyses will be stored within their own data base - some may just remain there. Police have a strong belief that knowledge should be produced and stored in case it will help some future investigation (Ericson and Haggerty, 1997, 319-20); the principle that 'we do not know today what we will need tomorrow' (Merseyside Police, 1983) can also mean that there is insufficient weeding of redundant material. Factors such as a mistrust of outsiders or a fear of compromising sources have sometimes meant that the simple accumulation of data has dominated the intelligence process.

If the product is disseminated to managers responsible for targeting decisions (↑10 in Figure 1.2) then it is 'feedback' - it completes the 'cycle' and may have some impact on a new round of targeting. Disseminating intelligence to other agencies (↑11) is the critical point at which power interacts most obviously with the information process. The objective of all cybernetic systems is to control their environment (Deutsch, 1966, 76). Of course, this is the primary objective of all police actions: dissemination is an important stage in the police's overall ability to take control of the criminal justice information process - in this task the police have the 'positional advantage' that the rest of the criminal justice system depends both on the police's accounts of 'criminal events' and of the procedures by which those accounts were produced (Ericson, 1981, 17-18).

Since it cannot be assumed that there is some consensus within an enforcement agency as to what the overall goals of the force should be (other than in the most general and rhetorical of terms), the influence of an intelligence unit within a police force might be measured by the extent to which it is able to recruit other units into sharing its perceptions of the definition of policing problems and potential solutions. In some cases it will set up a co-operative relationship with other units, in which case it will seek to trade by sharing, while in others it will establish a competitive relationship, in which it may seek to withhold information as a form of pressure or offer it conditional on reciprocal actions (cf. Wilsnack, 1980, 486-9).

A problem for CIUs is that, once intelligence is disseminated, they lose control of what meaning is given to the intelligence: if 'the meaning of facts changes as data move up or down the line in the organisation' (Manning & Hawkins, 1989, 149), 'intelligence' is even more prone to reinterpretation. Thus, when a unit disseminates its product, it cannot guarantee that it will be 'read' in the same way as it was written - recipients of the product may well subject it to their own evaluation and thereby challenge the ability of the intelligence unit to extend its influence by defining the terms of the debate about any issue. Therefore a crucial indicator of the success of a CIU will be the extent to which *its* perceptions of situations and likely future developments will be adopted by managers or operational personnel rather than, for example, their own 'recipe knowledge'. This is not just a question of 'selling the product' but involves a more fundamental challenge to define the terms within which organisational debate takes place.

When intelligence is disseminated the originating agency also loses control over what is done with it. For example, when allegations were made that the Chinese government planned to influence the 1996 US Presidential election by means of fund-raising for the Clinton-Gore campaign, investigations were initiated by Justice Department prosecutors, the FBI and congressional committees. Perhaps not surprisingly in an atmosphere where the Republican-dominated Congress was gunning for the Democratic President, accusations were made that information was not being shared as it should have been. Eventually the Justice Department's Inspector General investigated and reported that FBI officials had been at first reluctant to share 'raw intelligence' with Justice and congressional investigators. Later, the opposite was alleged: that Justice officials provided too much 'unverified' intelligence to Congress for fear of the political criticism that might follow if they did not do so! The Inspector General concluded that these failures did not amount to an intentional effort to interfere with any investigation (perish the thought!) but arose from institutional inertia, poor communication and a reluctance to share classified information for fear of compromising sources (*New York Times*, July 16, 1999). Clearly some of these represent failures that could be rectified but others indicate the structural issues at stake, specifically, that different, quite proper organisational interests will always prevent the 'seamless' dissemination of information, even if the atmosphere is less feverish than in Washington DC in the late 1990s.

In the quest for credibility, several factors are working in favour of the CIUs at the moment. Public sector managers, including the police, are now required to announce objectives and to establish performance indicators. The provision of strategic intelligence is clearly one way in which CIUs can help managers to establish these and thereby educate them (Faul, 1995). But if the intelligence is implicitly or explicitly critical of existing policies then it carries higher political risk:

> ...if an intelligence unit supervisor seeks to pursue a course of action, argues for a change in policy, or represents a questioning of existing practices in favor of those that are innovative, which challenges conventional wisdom, he must be prepared for the political consequences. Given this inherent contradiction, it is little wonder that most police executives will opt for the status quo, continuing a policy or practice that is of questionable or marginal utility. (Martens, 1990, 7)

Second, there are indications that, compared with twenty years ago, CIUs are much more conscious of the need to 'market their product' and, as we saw above in chapter seven, are having some success in setting up a virtuous circle in which operational officers are more likely to submit information in return for CIU assistance. There was a general recognition among interviewees in this study that they need to sell their product to often sceptical outsiders, for example, specific target packages developed by FIB in North Wales would be sent to the appropriate specialist squad or division and the FIB provided everyone in the force with a brief and glossy leaflet listing the things they could do. At divisional level intelligence staff use posters and briefings, emphasising the visual, to incorporate operational officers more into the process.

Barton and Evans examined what Merseyside police officers thought of the quality, relevance and amount of intelligence they received.

Table 9.1   Police views of quality of intelligence

	very good	good	aver-age	poor	very poor	no answer	total
number	23	101	105	33	12	16	290
%	7.9	34.8	36.2	11.4	4.1	5.5	99.9

Table 9.2   Police views of relevance of intelligence

	relevant	average	irrelevant	no answer	total
number	119	144	11	16	290
%	41.0	49.7	3.8	5.5	100

Table 9.3   Police views of amount of intelligence

	too much	right amount	not enough	no answer	total
number	27	97	151	15	290
%	9.3	33.4	52.1	5.2	100

Clearly, there are few doubts about the relevance of the intelligence that is being provided but more about its quality (15.5% saying it is poor or very poor) and especially the amount (52.1% saying it is not enough). Putting these into context, the authors of the research noted that LIOs in some cases were swamped with information (Barton & Evans, 1999, 30) and given, as we have seen, that much of their time can be taken up with simply inputting information, it is hardly surprising that they have limited time for dissemination. Yet LIOs are seen as the most useful formal source of information (see Table 7.4, p.167). The complaint about quantity may well represent the 'empiricist' fallacy that bedevils the intelligence process: 'any problem may be solved by obtaining more information'. Police are as prone as anyone to the belief that some additional information will provide the solution rather than subjecting what is already known to careful analysis.

A third factor is the evidence that CIU analyses have led directly to some legal changes, examples mentioned in Canada included asset forfeiture, changes in immigration regulations and amendments to 'name-change' procedures that now link new names with prior criminal histories (Engstad, 1995). Finally, the generally higher profile of 'organised crime' that has followed the collapse of the Cold War works to the advantage of CIUs since the forms of proactive policing required are those most in need of intelligence techniques.

On the other hand, difficulties remain. In general, the further intelligence has to 'travel' the more likely it is to get lost in the 'voids' that exist between different specialisms or 'levels' within the police (as represented in Figure 2.3, p.55). As we have seen, even at local level packages are more likely to be acted upon by dedicated 'proactive' teams than reactive squads working in traditional CID ways and others saw the inability to 'action' packages as a big problem. Silverman argues that the success of Compstat in New York results precisely from 'the successful convergence of information and decisions' especially at precinct level when specialist and generalist squads merged (1999, 181-4).

In North Wales intelligence developed locally on 'level one' targets will be sent to a proactive team if possible; intelligence on 'level two' - cross-border - targets will be sent to the FIB. Once jobs have to be sent up to force level then they are in competition for attention and resources with others and some (unknown) number are lost. At the next level, North Wales experience frequent frustration since jobs that they felt needed the resources of the then RCS could rarely compete (in terms of the points system) with those emanating from Merseyside and Greater Manchester. Conversely, where an operation failed to meet the RCS threshold and yet the Squad had what it considered to be good information then it would pass it on to a local force - if they were trusted (Nicholls, 1996).

Although CIUs may be able to sell their product to managers concerned with establishing objectives and so on, the nature of the crime problems particularly dealt with by CIUs are frequently 'low-visibility' (for example, drugs, prostitution, smuggling, money laundering) and therefore senior managers and policy makers may well respond more readily to public and media concerns about more 'high-visibility' street crime. This partly explains the fact that CIUs find it so difficult to be precise about their targeting priorities, as we saw above. They may believe that certain crimes are more significant in terms of 'damage' but, in order to safeguard their position, they have to be seen to respond to more immediate public and political concerns even if the crimes concerned are relatively less serious.

Also, the process of re-education into the ways of intelligence-led policing is slow: police officers remain often sceptical of the contribution of civilian analysts, especially if the police rotated into management positions in CIUs retain the narrow view of intelligence as serving only operational purposes (Fahlman, 1995). Many rank and file operational police retain the sub-cultural scepticism of 'intellectual' endeavours such as the work of the CIU. It may well be the case that intelligence work is coming to be seen as a desirable stepping stone in the career of an ambitious police officer (Dorn et al, 1992, 148), but the need for all to have some training in the value of intelligence is not being met. Normally, the work of CIUs is only dealt with in specialist and not in basic training. In the RCMP even the course for executive managers had no input from the Criminal Intelligence Directorate (Fahlman, 1995).

In choosing what to produce, CIUs can be caught in something of a cleft stick: they need to provide accurate and timely intelligence for operational purposes in order to encourage a flow of information back from those on the ground, but also want develop a more strategic view of crime in order to persuade senior managers and policy makers of their value at a time of budget restraint. There are a number of different forms in which intelligence may be disseminated. Peterson describes 37 different products (1995, 29-59) from assessments to warnings. The RCMP identifies, for example, the following: the Criminal Intelligence Brief prepared for senior officers and ministers relating some important intelligence development, a kind of 'warning' intelligence; the Strategic Intelligence Assessment identifying medium and long term developments for the Steering Committees that establish targets; Intelligence Estimates providing 'all-source' intelligence on particular areas for various clients, for example, the National Drug Intelligence Estimate; and Tactical Analytical Reports which provide charts and profiles for operational investigations (RCMP, 1995, 11).

New York agencies are more likely to report to political executives. This might be essentially tactical, for example, a report requested on some specific company (Haskins, 1999) or a report on a specific investigation that has produced recommendations for changes in policy, practice or regulations (Burke, 1999). The head of the NYPD Intelligence Division gives an oral daily intelligence briefing to the Commissioner and other senior officers (Oates, 1999). From time to time longer-term and often multi-agency projects will result in the preparation of more strategic assessments that may have some future policy impact and, because they are published, are clearly aimed to some extent at trying to influence public views in line with the law enforcement community's perception of threats. Examples of this would be the New York State OCTF Report to Governor Cuomo in 1989 on *Corruption and Racketeering in the New York City Construction Industry* and *An Analysis of Russian-Émigré Crime in the Tri-State Region* produced by OCTF, the New York State Commission of Investigation and the New Jersey State Commission of Investigation in 1996.

Dissemination and action

Finally, what do police do with the intelligence, if any, that is developed? What is the relationship of the intelligence product to organisational action? To what extent is the intelligence unit integrated with investigative or enforcement units? It might be hypothesised that the less the unit is integrated, the greater is the chance that its intelligence will not be translated effectively into action; on the other hand, integration might lead to the swamping of analysis by immediate investigative demands.

In its first years of operation NCIS worked on the principle of developing intelligence 'packages' regarding specific targets that they would then distribute for implementation to individual police forces and the RCS. These were judged to be of good quality but were unsuccessful because they were not 'timely' and it was unclear what 'value' was being added by NCIS to information often provided by a regional squad or police force. The approach was changed to one in which joint collection plans involving interested agencies were developed (Clay, 1997). Such strategic assessments as NCIS prepares will be passed to interested government departments, police and the Joint Intelligence Committee, whose weekly meetings are attended by the NCIS Director General.

'Timeliness' of dissemination is an important factor at all levels of policing. In Merseyside Barton and Evans found a number of factors leading uniformed constables to feel somewhat uninvolved in the drive towards intelligence-led policing and one of these problems was a perceived lack of support from Intelligence Co-ordinators, for example, that information provided to them on Fridays might not be disseminated until the following week and be too late (1999, 32).

If the intelligence does not simply enter the organisational store, it has three main alternative destinations. The first possibility is that it will be fed into some multi-agency policing process in which the primary objective is future-oriented crime prevention rather than some immediate investigative or prosecutorial outcome. This is typically the objective of such tools as crime pattern analysis and the identification of 'hot-spots' that may well require the action of authorities other than the police (Read and Oldfield, 1995 discuss the implementation of these in the UK). In some cases surveillance can be established that might result in the arrest of offenders but other measures may be taken regarding lighting or physical lay-out that aim to 'design out crime' and thus, in terms of this specific locality, disrupt it. This approach was adopted by the Merseyside Police area covering Liverpool city centre in which 'hot spots' rather than individual suspects were targeted:

> The primary goal of this area was to maintain a high police presence with the emphasis on disruption and crime prevention rather than arrests, and the uniformed patrols were particularly useful for this purpose. (Barton & Evans, 1999, 37)

The more general issue of disruption is discussed in more detail below. More holistic multi-agency strategies are now being encouraged in the UK *via* the concept of problem-oriented policing (Goldstein, 1990) and the programmes established under the Crime and Disorder Act 1998 that seek to advance information-sharing between police, local authorities and other agencies in the development of local community safety strategies. Clifford Shearing has argued that this kind of application of knowledge to problem-solving policing is the essence of current intelligence work but he exaggerates the abandonment of intelligence as 'a knowledge designed to promote deterrence through the identification of bandits' (1996, 294). Whether or not deterrence is successful, significant intelligence effort still goes into the identification and countering of 'bandits'.

This second possible destination for intelligence is into an investigation that then seeks to produce 'evidence' (rather than just further information) against specific suspects with a view to arrest, charge and prosecution. This requires a delicate process by which 'intelligence' that is not admissible in court (or which police do not wish to disclose for fear of 'burning' sources) can be converted into admissible 'evidence'. There are a number of techniques by which information that has already been obtained illegally will be 'washed' in order to obtain it in a form that will be admissible in court (Marx, 1988, 153), for example, undercover officers will be inserted into 'buy-busts' so that informants do not have to give evidence or warrants will be obtained so that the product of searches is admissible. In the Netherlands, police would conduct 'looking-in' operations (covert and unauthorised break-ins) in order to see if it would be worth obtaining a search warrant (Field & Pelser, 1998, 4). Other ways in which 'grey information' might be 'laundered' include using it in interrogation so that the pressure on a suspect to confess is increased by the apparent omniscience of the interrogators (Hoogenboom, 1991, 27).

If and when matters come to trial then the legality of the methods by which evidence has been obtained can be a central matter in dispute (Field & Jörg, 1998, 332-45 provide a comparison of the judicial regulation of covert policing in the Netherlands and the UK). It is not surprising that practitioners wish to keep these secret but it is crucial, given the recent role of non-disclosure in contributing to miscarriages and the problems of using informant evidence, that the defence is able to challenge. Both in the UK (*R. v Ward*, 1993) and Canada (*R. v Stinchcombe*, 1991) court decisions had the effect of increasing significantly the documentation that the state had to provide to the defence in cases being taken to trial. These cases had various impacts on police, including a significant increase in paperwork and, in Canada, the effect of taking the police nearer to the US model of arresting later in the investigative process (Ericson and Haggerty, 1997, 325-31) compared with practice in the UK. In the UK an energetic police lobby led to the Criminal Procedure and Investigations Act 1996 that clearly returned the balance of advantage to the prosecution (cf. JUSTICE, 1998, 62-7). As we saw in chapter eight, serious doubts have already been expressed at the fairness of these rules and their compatibility with the Human Rights Act is sure to be tested in due course.

Another factor now affecting the translation of intelligence into evidence is the growing role of security intelligence agencies in law enforcement: One of the major reasons why 'security' and 'law enforcement' intelligence have been seen as *essentially* different was that the latter was concerned with the arrest and prosecution of criminals while the former was not; rather it was concerned with the neutralisation of security threats. It is possible to exaggerate the significance of this difference; for example, since the days of Rowan and Mayne, the stated object of police was 'the *prevention* and detection of crime...' but there is a limited sense in which this distinction was real. While police regarded arrests and convictions as a 'result'; for security intelligence personnel, a court appearance was to be avoided if possible and a prosecution might reflect a failure to turn someone into an informer. Now, their increased involvement in law enforcement has required some increase in court appearances. Normally, they have appeared behind screens and given evidence anonymously (eg. JUSTICE, 1998, 67-9).

Such practices raise issues of principle and practice for the integrity of the criminal justice process and the rapidly-increasing significance of disruption as a law enforcement technique (the 'third way'?) raises even more. It has been police practice for longer in Germany and the Netherlands (Field & Pelser, 1998, 11-12) but has attracted relatively little comment in the UK despite being practised increasingly by law enforcement agencies. For example, one of the Home Secretary's objectives for the National Crime Squad in 1999/2000 is:

> to improve...the quality of operations leading to the arrest and prosecution of individuals or the dismantling or disruption of criminal enterprises, engaged in serious and organised crime within or which impacts on the UK. (NCS, 1999, 7)

But the objective regarding serious and organised crime set by the NCS Authority itself actually refers to 'targeting and disrupting', not arrest and prosecution, although the number of arrests is one of the performance indicators (NCS, 1999, 8). Similarly, the key objective of UK Customs regarding drugs is 'to improve the effectiveness of drugs enforcement, with particular emphasis on commercial smuggling and disrupting the international supply of drugs'; and, in these days of performance measurement, there is a consequent measure of the number of illegal organisations disrupted - 130 claimed in 1997-98 (Select Committee on Public Accounts, 1999). Disruption was not referred to in the North Wales policy document published in 1996 but several interviewees acknowledged its utility.

Disruption has become more significant because it enables police to avoid the costs and uncertain outcomes of the criminal justice process while it fits in well with the idea of intelligence-led policing. In the 'classical' model of criminal investigation police react to the commission of a crime by an investigation at the end of which some people are charged with an offence. Disruption makes no sense within this model because investigation is, by definition, after the event. The new model of criminal investigation is proactive and takes two main forms that may actually overlap. First, 'known' criminals will be targeted and information will be gathered as to their activities. These *may* result in the gathering of sufficient *evidence* of actual crimes committed that arrest and prosecution will be the outcome; however, this may not be possible because, for example, of the amount and length of surveillance (i.e. cost) required whereas the criminal enterprise can be disrupted relatively quickly and cheaply.

For example, it is suggested that key players within some criminal organisations protect themselves so well from direct involvement that they have become effectively 'untouchable' and it is in these circumstances that police consider the alternative of disruption. The police feel that it would be useful to have more powers that could be used in this way, for example, HMIC strongly recommend bringing in civil proceedings for asset forfeiture (1997b, 4.21-24). In general, the argument goes, if police are unable for whatever reason to generate the necessary evidence upon which to arrest and prosecute for criminal activities, then the next best thing in terms of public safety is to prevent a specific crime occurring or by seizing drugs or illegal goods to interrupt the process by which illicit profits are generated. Indeed, if an entire organisation or network is disrupted then that is arguably more effective than the arrest of a few people who will quickly be replaced. Against this, others argue strongly that disruptive strategies are counter-productive, unless used as a means to the end of getting closer to a major player, since they will merely divert the criminal entrepreneur into some other field of activity.

Public-private sharing

Within the context of developing public-private intelligence networks a key issue is the extent to which the product of public agencies is disseminated to private agencies (cf. Hoogenboom, 1991). Customs, for example, deals extensively with the private sector in terms of its revenue activities and potentially traders are a major source of information. One interviewee made it clear that UK Customs regard dealings with 'the trade' as an one-way street for information in which information cannot be passed back (Parker,

1997) but it is difficult to see that this formal position can always be maintained. For example, traders may well view provide information to Customs on this non-reciprocal basis if their enforcement activity is complementary to their own associations in combating fraud, but in other circumstances traders are likely to expect some more direct reciprocity. At force level, police guidelines regarding dealing with other agencies were limited, though formal contacts with other public agencies in such areas as child abuse and with probation regarding the release of those considered 'dangerous' were extensive (Williams, 1997). These relationships will no doubt be developed further within the context of the multi-agency partnerships mandated by the Crime and Disorder Act 1998. Contacts with private sector agencies tended to be informal and personal; formal relations were limited, for example, to briefing local shops about relevant threats.

Other examples of mutual aid between public and private policing sectors emerged in the research, for example, the RCMP sought information from major credit card companies and the Canadian Banking Association, and, in return, provided assessments regarding credit card fraud (Fahlman, 1995). CISO also did a strategic report on credit card fraud for the Banking Association who, in turn, had provided financial support for CISO (Faul, 1995). In British Columbia CLEU advised the Insurance Corporation on ways to change their business practices after an investigation into criminal gangs involved in car theft and insurance fraud (Engstad, 1995). Ericson and Haggerty illustrate further just how extensive is the two-way street in information between public and private police (1997, 203-4).

This kind of contact can only become more important: first, because of the developing networks between cash-strapped public police and the faster-growing private security sector (Ericson and Haggerty, 1997, 167-72; Johnston, 2000, 163-75) and, second, because of the increasing inter-penetration of organised crime with 'legitimate business' (Beare, 1996, 221), or, in other words, between 'underworld' and 'upperworld' (e.g. van Duyne, 1996, 342). In the UK the interweaving of public and private policing is slightly more discrete than in Canada, for example, the author listened to a local crime prevention officer at his local neighbourhood watch meeting bemoaning the fact that he was prohibited from recommending particular commercial suppliers of security equipment to householders, thus leaving them at the mercy of 'cowboys'. However, there is a deal of ambiguity in official pronouncements regarding the dissemination of information by police. At one level police are warned as to the potentially damaging consequences of unauthorised disclosures to the press culminating in the recommendation that 'the Service should embrace more closely a culture of confidentiality, whilst at the same time freely sharing information with partnership agencies' (HMIC, 1999b, paras.6.15-17). Installing a structure that could realistically achieve both these ends, if even theoretically possible, would be so bureaucratic as to encourage further the very informal exchanges of information that are the lifeblood of public-private security networks.

But the Code of Practice on dissemination recently published in the UK does not attempt this; there is nothing in the Code to inhibit dissemination within the public sector since officers may authorise themselves without the requirement of an audit trail unless the dissemination is computerised (NCIS, 1999e, 3.6). In the case of dissemination to private agencies an Inspector's authorisation is required and a record will be kept of the material sent, to whom and for what purpose (3.10). Police inspectors may also authorise dissemination to foreign law enforcement agencies within the European Economic Area (EEA - the EU countries plus Iceland, Liechtenstein and Norway). A Superintendent's authorisation is required for dissemination outside that area. Within the EEA intelligence exchanges take place within the context of Conventions such as those on Covert Operations and Co-operation between Customs Administrations and if the standard grounds apply then dissemination may take place if it is 'deemed a necessary means to effect the desired results' (4.4). Outside the EEA, dissemination may take place for the prevention or detection of international crime and terrorism. The Data Protection Act is quoted to the effect that personal data may not be transferred outside the EEA if that country has inadequate protection for the rights and freedoms of data subjects unless 'a substantial public interest would be served by the transfer' (4.3). It is worrying that the complex judgements that might well be required in such cases can be resolved at the relatively low level of Superintendent. In all cases of dissemination abroad a record will be kept of the material, the addressee, any relevant restrictions on further dissemination if the material is confidential, the objective of the dissemination and the risk assessment in cases of countries outside the EEA. The issue of disseminating intelligence material to transnational policing agencies is not explicitly dealt with; presumably Europol is covered by the EEA provisions but what of Interpol that is not referred to in the Code - is it an 'agency' and, if so, is it within or without EEA?

Conclusion

Criminal intelligence analysis is a young specialism within law enforcement and, despite its growth, remains vulnerable. This is a structural problem that will remain inherent within the organisation of policing unless there is a significant cultural shift. Analysts, in addition to being generally younger, better educated and more female than most police, find themselves between the rock of operational personnel and the hard place of ill-informed managers. Operational personnel may distrust intelligence as an alternative form of knowledge to the extent that they do not perceive it as immediately useful to them and if they see it being used by managers to shift policy in directions they do not like. Meanwhile, most police managers will have risen to their current eminence with little or no understanding of intelligence, for example, they may not appreciate the significance of the difference between 'information' and 'intelligence' and be cynical or sceptical as to the value of intelligence. The HMIC Report on *Policing with Intelligence*, in part intended to disseminate best practice, itself betrays very poor understanding in places. For example, in a section headed 'Analysis' it describes the 4x4 system (see Figure 7.1, p.163) for assessing the credibility of information gathered; this is certainly a pre-condition for analysis but does not in any way constitute analysis as such (1997b, 3.11). The same report also betrays too much faith in the technological fix: that as more officers can input more data and search more databases then there is progress! (1997, 2.28-29). At the very least, what must be done to try to address the isolation of the analytical function is to train not just intelligence officers and analysts but also to train non-intelligence operational personnel and managers into some realistic view of the possibilities and limits of intelligence (Dintino and Martens, 1983, 123-4; Peterson, 1999, 7-8).

Within intelligence units, there are other problems that must be guarded against, for example, 'mirror-imaging', that is, the tendency of analysts to assume that targets behave in similar ways to themselves. Clearly, the greater cultural distance between analysts and targets the greater is the danger that this will lead to serious errors of judgement, for example, the tendency of those who work within hierarchically-organised bureaucracies to assume that criminal enterprises that have been targeted will be organised similarly. It can be argued that the longevity of the 'Mafia myth' results in part from this factor. If, as has been argued here, the target of law enforcement intelligence are networks (incorporating individual and group enterprises of great flexibility) operating within illegal markets then it reinforces Sparrow's conclusion that analyst themselves must be deployed much more flexibly, and closer to the network form:

> …information management resources (analysts and equipment) will need to be deployed flexibly, on a project basis, in teams of many different sizes. For analysts that means a major change in professional lifestyle. They are accustomed to working roles or positions, not projects. They are accustomed to their own place within a hierarchical management structure, not to working alone on a small project one day and as part of a sizeable team on a major program the next. (1994, 110)

Resistance to departing from the hierarchical form will be considerable: managers do not like losing control and will doubtless cite the need for security in intelligence units. This is indeed an issue, the sensitivity of the information gathered and the possible compromise to future investigations mean that there must be security consciousness including the possibility of penetration from outside in the form of infiltration or disinformation. This may be less relevant as a factor for police than for the counter-intelligence sections of national security agencies, but if transnational criminal organisations are becoming as sophisticated as some suggest then the possibility clearly exists. Certainly there is widespread belief among those policing major crime that the more sophisticated criminal enterprises use such tactics.

Finally there are the considerable privacy implications of the current technological advances in the relating and matching of data. Seen by both public and private agencies as essential tools in countering fraud, it is clear that the technology is far in advance of the legal frameworks within which it takes place. For example, the Social Security (Fraud) Act 1997 enabled large data-matching exercises between several government and local authority departments without safeguards considered essential by, among others, the Data Protection Registrar (JUSTICE, 1998, 98-99). Four of the five codes of practice issued by ACPO and Customs in 1999 refer to various means of gathering information (see chapters seven and eight) and the fifth refers to the recording and dissemination of intelligence. This leaves a significant gap: the processing and analysis of information is not covered at all in the codes and nor is it covered by the Data Protection Act which does not regulate law enforcement in the collation and manipulation of the information they have stored.

It is clear that the effectiveness with which intelligence is disseminated is a key indicator of the value of a CIU but this will be difficult to measure in terms solely of formal products such as briefings because of the importance of informal sharing and networking within and between agencies. The preference of practitioners to maximise the flexibility and fluidity of such networks is part of the explanation for the growing popularity of disruption as a law enforcement tactic. The criminal justice process is centrally concerned with the propriety of official evidence-gathering methods and requires audit trails that do not sit happily with informal exchanges. To the extent that disruption permits the translation of intelligence into action without any need to account publicly for its provenance, then this growth in popularity is understandable but, by the same token, it poses the severest challenge to traditional notions of accountability.

10 Rounding Up the Usual Suspects?

Introduction

Despite the fact that the arrival of the new Millennium passed without the general chaos and disorder predicted by some, the beginning of the twenty-first century has seen no reduction in levels of official concern that organised crime and/or terrorism must be countered by yet more resources, powers and effort by law enforcement and security intelligence agencies. For example, in Canada, the Government announced at the end of February 2000 that extra funding would go to the CSIS, RCMP, Immigration and Customs, each of those having suffered budget cutbacks during the 1990s (*Ottawa Citizen*, February 29, 2000). The immediate catalyst for these announced increases was the 'security scare' at the US-Canada border at Christmas 1999 (see chapter two above) that gave rise to a bout of criticism of the alleged laxity of Canadian border security. Much of this came from the US but the support it had from some Canadian commentators led to suspicion that the cross-border law enforcement community was orchestrating a campaign for extra funding from both governments.

Meanwhile, in the UK, growing official concern that organised crime might be running 'out of control' led to a Home Office review of the various agencies involved in which each set out its strengths and proposals for the future. Reportedly, one possible outcome of the review will be to construct unified national intelligence and operational units, the former from merging NCIS, Customs National Intelligence Division and MI5 officers currently allocated to 'serious crime' and the latter from combining Customs Investigation Division and the National Crime Squad. MI6 and GCHQ would remain separate though both seeing opportunities to develop their crime work (*Independent*, February 26, 2000). Such a re-arrangement might well rationalise the current fragmented national law enforcement to some extent but it would probably not reduce the overall fragmentation nearly as much as might be imagined. Unless, for example, the mandate of the Security Service was restored to the *status quo ante* 1996 (that is 'serious crime' was removed) and that for Customs was amended, the fact that those two agencies plus MI6 and GCHQ would still be independent would mean that the re-organisation would be little more than a merging of NCIS and NCS. Arguably, these should have been merged from the outset and were not only because of the opposition to a 'national police' from chief constables. If the UK does get its own 'FBI', it will be, as in the US, just one among a number of law enforcement players on the national stage.

Is intelligence-led policing making a difference?

At the end of chapter one a number of questions were posed as to whether or not ILP was making a difference to policing, for example, are investigations being more directed by intelligence priorities compared with traditional reactive strategies, is policing being transformed by a combination of ILP as a strategy and the associated technology? Things are certainly changing but they are doing so unevenly. What is clearly changing is the language and rhetoric with which senior police managers, especially in the UK, have embraced the new techniques. What is less clear is how far this is matched by actual change within enforcement agencies. The extent to which any agency embraces ILP depends on a range of factors including size, mandate, resources and managerial commitment. Larger agencies and those with primarily investigative mandates are more likely to sustain separate intelligence units which, given adequate resources may be supported with appropriate ICT but even where this is the case, extending ILP throughout agencies as an integrating strategy is much rarer.

Is there evidence that ILP is overcoming the rank-and-file resistance to innovation that is traditional in police agencies? Ericson and Haggerty argue that the formats required by the new technology and accountability demands can determine more precisely the information provided by police and thus affect how they think and act (1997, 322). These authors are in no doubt that things are changing:

> Communication technologies...radically alter the structure of the police organisation by levelling hierarchies, blurring traditional divisions of labour, dispersing supervisory capacities and limiting individual discretion (1997, 388),

and

> Our data indicate that knowledge, power, authority and hierarchy are being transformed in police organisations by the infusion of communication technologies. (1997, 411)

Yet, they also acknowledge that there is nothing deterministic about these shifts, that they will not necessarily take place in the same way everywhere and face a variety of problems and resistance from within police organisations (1997, 412-18). Maguire's research into the development of ILP in the UK leads him to conclude similarly that 'a definite shift is detectable in both the language and practice of crime-related aspects of policing' (2000, 332-33). This study has also identified some of these shifts, for example, reorganisations at local level in line with the crime management model and greater explicit use of targeting as a basis for the deployment of investigative resources. But we have seen also that the implementation of such changes has been uneven, partly they depend on the will of supervisory officers to adapt but everywhere they are subject to external events and demands that can derail even the most carefully thought out proactive strategy.

But is ILP working in terms of the general objective of 'reducing crime'? Based on the idea that police investigatory effort should concentrate on known offenders rather than reacting to individual criminal incidents, the adoption of ILP did provide a potential test of the common belief among police officers that they 'know' who are the local burglars and car thieves; the problem is to gather enough evidence to convict them. At first glance, the adoption of intelligence-led policing from 1993 onwards appeared to have an immediate effect. Recorded crime, having risen inexorably since the early 1950s fell from 1993 until 1998 (*The Guardian*, October 13, 1999, 4). However, such a conclusion would be premature. Even if recorded crime statistics are accepted as an accurate measure of the extent of crime, other factors possibly contributing to a fall would need to be taken into account. Police strategies are only one, and probably not a very important one, among a range of factors that might contribute to levels of crime. But recorded crime figures are not an accurate measure - it is generally accepted that a better measure of the extent of crime is obtained from crime surveys. The British Crime Survey showed the level of crime continuing to increase through to 1995 but then it also showed a fall (Barclay & Tavares, 1999, 7).

Arguably, by allowing time for the new methods to actually start having some effect, this might seem actually to strengthen the argument as to the impact of ILP but, again, this would be a premature conclusion. It has been shown, for example, that comparing research data on repeat offenders (the statistical basis for targeting 'known offenders') with available police resources shows that it is quite impossible for police to target systematically more than a small proportion of 'repeat offenders'. This does not mean that there is no point in police seeking to target the most prolific offenders (and thereby reduce the social damage they cause) but it does mean that even 'targeted' policing will impact directly only on a small proportion of crime (Heaton, 2000).

Careful study of the impact of the new techniques in specific areas is required before firm conclusions can be reached. In particular, great caution must be exercised in making use of statistics collected by police themselves in assessing the effectiveness or otherwise of police strategies. Decisions made by police at key points in the investigative process as to whether incidents will be recorded as 'crime' and, if so, which one, have always meant that figures could be massaged in line with whatever effect was needed at the time. The application of 'performance indicators' to police has complicated the picture further, for example, providing further encouragement to massage statistics and arguably distorting police practices towards those outcomes capable of being measured quantitatively regardless of their significance for the quality of policing.

The Home Office established research studies into several aspects of the new strategy, for example, a study of the adoption of the new crime management model in the early 1990s. This found that the time allowed for the evaluation was too short and that the speed of organisational innovation within the forces concerned made controlled comparisons impossible. It concluded, therefore that, in terms of reduction, the value of the model was 'not proven' (Amey et al, 1996, 32-3). There has also been research into the impact of problem-oriented policing, at the heart of which is the greater application of ILP - and this has produced mixed messages. Again the model is seen as being adapted differently by different forces and this uneven implementation prevented evaluators from reaching firm conclusions but they noted that implementing the model is more difficult than it appears (Leigh et al, 1998, vii). Further studies, one into problem-oriented policing on Merseyside and another into ILP did reach one similar central finding - that both can prosper only in the context of fully integrated applications within forces that require comprehensive cultural and organisational change driven by committed and sensitive managers, though neither examined specifically the impact of the initiatives on crime (Barton & Evans, 1999, 44-46, Maguire & John, 1995, 54-55). Silverman has recently argued that NYPD has achieved just such a shift and major impact on crime since 1994 with its systematic application of Compstat (1999, 179-204) though it must be pointed out that similar reductions have occurred in other American cities without the same police strategies. Any systematic evaluation of ILP would have to examine the investigative process from beginning (what police record from incoming calls and what then happens to those 'incidents') to end (the disposition of cleared cases in terms of charge, caution or NFA), deploying both an audit of the documentary trail and observations of and interviews with officers.

Surveillance, intelligence and the 'geology' of policing

At the end of the first chapter it was suggested that the fragmented network characterising contemporary law enforcement intelligence structures might best be represented by a 'geological' metaphor of rock layers, fractures, fissures and so on. As a specific example of the more general problems of contemporary governance this causes problems for practitioners but it concerns also anyone who perceives the state as having some potential for the protection of collective or public interests over those of the individual and the private. For example, Hirst and Thompson criticise the more extreme versions of the 'globalisation' thesis for their abandonment of the idea that states can regulate markets. Acknowledging that there has been a shift to 'governance', they argue:

> ...governing powers cannot simply proliferate and compete. The different levels and functions of governance need to be tied together in a division of control that sustains the division of labour. If this does not happen then the unscrupulous can exploit and the unlucky can fall into the 'gaps' between different agencies and dimensions of governance. The governing powers (international, national, regional) need to be 'sutured' together into a relatively well-integrated system. If this does not happen then the gaps will lead to the corrosion of governance at every level. (1996, 184)

One can imagine many law enforcement practitioners saying 'amen' to that. However, as we have seen, the problem for practitioners is that the 'governing powers' *are* proliferating and often competing - initiatives such as interagency task groups or co-ordinating agencies that are intended as 'bridges' (cf. Hirst and Thompson's 'sutures') across 'voids' can quickly develop their own organisational interests so that they may actually develop into fresh 'islands'. Perhaps there is too much concern with organisational change, after all, is it not precisely the advantage of ICT that it can facilitate information exchange and co-ordination remotely without requiring people to occupy the same office space? Certainly, this is so, but as we saw in chapter seven, there remain considerable barriers to what idealists would consider the effective transmission of intelligence between remote points if the essential elements of trust and reciprocity are lacking.

But where Hirst and Thompson are concerned that the weak will be exploited if the gaps between governing levels remain, others are concerned that it is, to the contrary, precisely the rapid closure of these gaps that poses the greatest threat to the weak. For example, Mathieson argues:

> ...there is a tendency towards convergence and integration between the various registration and surveillance systems - established or in the making - in Europe...Though there are obstacles and conflicting interests involved, the tendency has a momentum on the political as well as the policing levels...On the horizon, we may envisage the contours of a vast, increasingly integrated multinational registration and surveillance system, with information floating more or less freely between subsystems, at any time covering large population groups. (1999, 29)

So, is Mathieson, as it were, reassuring Hirst and Thompson that what they want is coming about? No, because they see nation-states as still the primary source of law and legitimate regulation for whatever variety of governing bodies emerges (1996, 191-94) and therefore the only potential check on an otherwise unregulated global economic jungle. Mathieson, on the other hand, is concerned that the growing transnational integration of surveillance will lead to profoundly undemocratic outcomes, including the illegal surveillance of those viewed vaguely as posing possible future threats to security and order (1999, 29-30). Can these accounts be reconciled?

Only if it is acknowledged that surveillance is not necessarily zero-sum. In other words, surveillance in all its forms may be the antithesis of privacy and the widespread technologies of surveillance now available constitute an invasion of personal privacy unimagined in previous times but surveillance is not therefore always negative. Great damage is wrought by people acting within the private domain, from corporate managers in offices and suites to violent men in homes and streets, and it cannot be argued that there are no circumstances under which it would be proper to surveill their behaviour. Since there is clearly no way in which these technologies can be disinvented, and, note, they emanate increasingly from private corporate interests, not the state (Castells, 1997, 300-01), the problem then is to regulate their deployment.

Before we consider how that might be, there are other specific issues relating to the deployment of law enforcement intelligence. We are currently confronted by two very different images of the criminal justice process, one is of an ever-increasing totality of private and public surveillance leading to our entrapment in a growing web of social control. An obvious example of this is the inexorable growth (in the UK) of CCTV schemes in public places and of the sophistication by which data can be processed, transferred and deployed in the 'prevention' of crime and disorder. This is all complemented by the steady extension of the criteria for surveillance from generally-acceptable notions of 'violence' or 'crime' to much more inclusive and debatable concepts such as 'alarm, harassment and distress' (Public Order Act, 1986, s.5) and 'community safety' (Crime and Disorder Act 1998). The result is an increase in the information available to authorities so massive that ever greater social control is facilitated.

The other image is of a criminal justice process that makes, at best (or worst, depending on point of view), the most marginal of impacts on social life. As we have seen, it is estimated that just 3% of serious offences in the UK result in a conviction or caution and, of those convicted for any offence, only 7% were sentenced to immediate custody (Barclay & Tavares, 1999, 29, 41). If there is one constant public call in the UK it is for 'more coppers on the beat'; regardless of the minimal impact of patrol officers on the prevention and detection of crime, their presence is clearly something that people in general find reassuring. But the numbers of police remain fairly constant and organisational pressures such as training, leave, administration and specialisation ensure that a very small number of police are actually available for patrol at any one time. As we saw in chapter six about half of crime that is reported to police locally is 'screened out' from further investigation and, indeed, ILP has sought to make a virtue of proactively targeting known offenders rather than responding to public reports of crime.

The centrality of targeting

It is the process of targeting that goes some way towards reconciling these widely divergent images. For the entirely understandable reason that criminal prosecutions can only be mounted against named individuals, practitioners are concerned primarily with identifying and gathering information about specific criminal actors. Since the police cannot identify, let alone gather information on, all of them, the process by which some are selected for attention over others is crucial to understanding contemporary law enforcement. This is not a rational process, based on some environmental scan in order to identify those causing the greatest damage, but is an organisational-political process within or between agencies. As we saw in chapter six, some agencies have made attempts to systematise the process but those schema provide a rationalisation of and context for discussions about targeting rather than acting in any 'objective' way.

These discussions take place within a decision environment that sees the confluence of a number of strands including organisational culture and beliefs, feedback and spin-offs from previous work, requests and pressures from other agencies and the level of general political rhetoric about specific crime problems. Sometimes these pressures may clash but their overall direction is clear. The requirement among politicians for some 'punitive display' leads inexorably to the identification of some criminal 'other' (Garland, 1996) as being responsible for crime. Since police in Canada and UK are not subject to political direction - there is probably greater direct input from political institutions into police targets in the US - they do not have to respond to that directly, and may make great play of their independence, but it cannot be denied that the expression of strong official ideas on targets has influence on police. This will be even more so, where, as often, ministers' own ideas are based on police briefings as to the nature of emerging crime 'problems', or, if not, the ideas of police themselves are similar. Research has shown the continuing significance of class, race and gender stereotyping in the police culture and this permeates the targeting process. Law enforcement officials everywhere seem unable to avoid the discussion of 'organised crime' targets in predominantly ethnic terms. This is highly damaging to the integrity of the law enforcement intelligence process. If, as they often claim, it is just a convenient shorthand for the persistence of specific crime problems within particular ethnic communities, then they have to realise that it is highly offensive to many people (and, arguably, could become illegal under the proposed changes to the UK Race Relations Act). However, to the extent that such stereotyping is reflected in actual targeting practices that go beyond shorthand figures of speech, then what is presented to the public as newly-rational 'intelligent' police practices is actually little more than traditional policing of the 'dangerous classes'.

This is not claimed to be an original insight - some police and other officials pointed out some time ago that crime strategies that target specific criminal actors or associations and measure their 'effectiveness' by resulting head-counts simply fail to get to serious grips with ongoing activities within criminal markets (e.g. Bersten, 1990; Pagano *et al*, 1981). There are efforts to escape from the short-term circularity of much law enforcement, for example, the Dutch Strategic Intelligence Research (SIR) system. Formally, this has three phases: first, a society scan, sometimes referred to as 'phenomenon research' in which a wide variety of data including open sources, interviews and state information would be drawn on to provide an overall picture of 'the problem' or sector of society perceived to be vulnerable to illegal activities, for example, the concentration in an area of particular illegal activities or an increase in the social damage caused by the operation of certain illegal markets. At this stage the research is more like a social science research project. At the second stage a sector scan studies the specific factors making the sector vulnerable including identifying more specific individuals and organisations. The insights gained at this stage are combined in the third stage with more specific law enforcement data in an organisation scan or reconnoitering investigation that might drive a normal police investigation or the development of a preventive strategy that uses forms of regulation other than traditional law enforcement ones. In practice, the system is more flexible than this suggests; investigators themselves may conduct a general scan of files at their disposal as a result of a specific tactical inquiry. For example, as a result of a specific investigation the model was followed in Amsterdam to identify recently-established currency exchange booths as a primary location for the laundering of drugs money and the issue was tackled by the increased regulation of the booths (Howe, 1997; Stelfox, 1999, 7).

The National Intelligence Model in the UK

Recently, there are more signs of similar thinking in the UK (e.g. Sutton, 1998) but much greater effort is still required to shift intelligence attention away from actors and towards damaging market activities and the crucial question of how regulatory activity itself impacts on the market. The most recent attempt to spread the intelligence gospel in the UK has been the promulgation by NCIS of the 'National Intelligence Model' in February 2000. It discusses both intelligence structures and processes and thus seeks to provide a handbook for the creation of an intelligence unit while also arguing for common procedures to be adopted to contribute to 'joined-up' law enforcement (NCIS, 2000, 7). The intended outcomes are said to be improved community safety, reduced crime rates and the control of criminality and disorder. The Model itself has four components: the tasking and co-ordinating process, 'intelligence products', 'knowledge products' and 'system products'. The first of these require regular meetings that will develop the 'control strategy' by means of both strategic and tactical tasking, set priorities for intelligence work and provide for the maintenance of the intelligence unit. The main intelligence products are strategic and tactical assessments; the latter includes the development of target profiles for either individuals or 'problems' such a hot-spots, crime series, and prevention measures (NCIS, 2000, 11-23).

So-called 'knowledge products' are the variety of local, regional and national rules and information that need to be learnt by staff in order to enable them to fulfill a role within the intelligence process. They include codes of practice, legislation, rules of access to secure systems and protocols with other agencies. 'System products' refer to the provision of, or access arrangements to the 'store' and open sources (cf. chapter seven), to facilities for acquiring new information and protective security (NCIS, 2000, 24-28). For those already familiar with the idea of the intelligence cycle, describing these two components as 'products' is confusing. They both describe crucial *processes* within the model but to call them 'products', by confusing means and ends, is simply inaccurate.

In the other two sections of the document, nine 'analytical techniques and products' (another confusing use of the term - what is wrong with describing them simply and accurately as 'techniques'?) are outlined (cf. discussion in chapter nine) and attention is given to the issue of the links between the different levels at which police operate. As we saw in chapter four, this has given rise to some controversy within the UK because of the perception at local force level that the regional and national agencies were moving away from their concerns to concentrate on transnational crime. It is suggested, correctly, that working across the levels is facilitated if the agencies concerned are all using the same model but the claim that 'the Model provides the information and decision making solution to what has been termed "the regional void" in law enforcement' (NCIS, 2000, 12) cannot be sustained. As we saw in chapter two, the regional level is just one of a number of 'voids' that exist between levels and, though the skilful application of techniques such as criminal business profiles may well *assist* in the process of determining priorities and targets at different levels, issues that go to the heart of organisational mandates and interests cannot be 'solved' technically.

The aims are certainly ambitious - to provide a model that can be applied to all law enforcement needs for both strategic and tactical purposes - yet its publication has to be understood in the context within which it has been produced: a somewhat marginalised NCIS seeking to convince police forces of both the value of ILP in general and NCIS in particular, specifically, that it has not lost sight of their problems as it operates increasingly in the transnational arena. Given this, the presentation ends weakly with a case study of how the model may be applied to 'an international problem' - stolen cars. If one were seeking to sell to a sometimes sceptical audience the value of an integrated intelligence model to tackling the damaging consequences of international illegal markets, it is hard to think of a poorer example (in terms of the social damage caused) than stolen cars!

Intelligence and the growth of 'informalism'

Whether we consider police intelligence at what the new UK Model terms level one (local) or level three (serious and organised crime - usually on a national or international scale) there is an important common element, that is, the extent to which newer strategies seek to *avoid* rather than *involve* the criminal justice process. As we have already noted, of itself this is nothing new since one of the original aims of the Metropolitan Police was the 'prevention' of crime *via* its patrol and associated surveillance activities and there are a series of structures within the criminal justice process that are geared to the avoidance of trials (Ashworth, 1998, 301-06). But more recent developments are indicative of the shift from 'sovereignty' to 'steering' that was discussed in the first chapter and are reflected in the clear trend towards greater use of both formal and informal means of dealing with crime problems outside of the criminal justice and penal systems (Maguire, 2000, 317). The immediate riposte to this statement might be that, if it is so, then how does one account for the upward trend in the numbers imprisoned in the UK since 1990? The number of persons sentenced to immediate custody in 1987 was 74,000, by 1990 it fell to 58,000 but rose to 93,000 by 1997. This was not because of any significant increase in the numbers of people being prosecuted in court; indeed, in 1987 this was 1.92million, rising to 2.05m by 1992 and then falling back to 1.86m in 1997. What does explain increased incarceration is the sentencing policy in courts - of those convicted for indictable (the more serious) offences, not only did the proportion sentenced to immediate custody increase from 14% in 1990 to 22% in 1997 but the average length of sentences given by Crown Courts increased 20-25% (Barclay & Tavares, 1999, 32, 44, 46).

The conclusion that it is the courts rather than the police or prosecutors who are responsible for increased incarceration is reinforced by Hillyard and Gordon's examination of arrest statistics. They show that during the period 1981-97 the number of people arrested by police has steadily increased, for example, during the period 1987-97 the numbers arrested increased from 1.4m to almost 2.0m while there was only slight fluctuation in those brought before the courts (1999, 508). This cannot be explained by changes in formal police cautioning - it increased slightly until 1993 and then decreased slightly; what has been happening is an increased use of arrest followed either by police taking no further action beyond, perhaps an informal caution or, if police lay charges, by the CPS discontinuing the case.

Without a detailed analysis it is impossible to determine the extent to which this growing 'informalism' might be seen as progressive or regressive. Much depends on precisely what motivates it and the destination of cases that do not proceed to court. For example, if those who are arrested while abusing alcohol or other drugs are diverted by police and/or other agencies to detoxification or treatment centres rather than being processed through courts and into prisons then the outcomes are potentially better for all concerned (e.g. Dorn, 1994). Similarly, the use of police cautioning, especially of juveniles has always been justified on the grounds that it seeks to divert people from the criminal justice system in the knowledge that the consequences for individuals of being caught up in the formal process are rarely positive. But while elements of these positive reasons for diversion remain, other motivations for current developments raise more troubling issues.

One element of current thinking does not amount to 'informalism' as such but does seek to ease the evidential burden on authorities by importing civil procedures and the lower burden of proof into law enforcement. We have seen a number of examples of this in this study, for example NYPD's civil enforcement initiative involving the deployment of nuisance and housing laws against specific sites of drug markets (Silverman, 1999, 125-45). A variation of this is the current move in the UK so that those suspected of involvement in organised crime and who are shown by financial investigations to have great wealth but no legitimate means of income will have the burden of proof shifted to them to prove the legitimacy of their wealth. This proposal would extend to crime in general the measures first incorporated into the Prevention of Terrorism Act 1989. Existing measures to confiscate the proceeds of crime are not seen as having been successful, restricted as they are mainly to drugs cases and requiring a criminal conviction (Goddard, 1999). While the reduction of the criminal burden of proof can be criticised in such cases because it would make it easier for the police and other agencies to 'get a result' against someone if they had failed to secure a conviction, at least it does preserve an element of transparency.

Although depriving people of their 'ill-gotten gains' is undoubtedly part of the motivation for recommending civil forfeiture procedures, it is highly tempting for states to seek such funds to compensate for their declining tax-base. However, there is another sense in which cost is a significant motivating factor for the shift towards informalism. Gathering information is not cheap, especially if it requires covert methods, but it can still be cheaper to gather enough information through surveillance in order to *disrupt* some element of a criminal enterprise than it is to gather *evidence* against one or more of the entrepreneurs that will, first, pass muster with the

CPS, specifically the so-called 51% rule that the evidence is more likely than not to lead to a conviction and, second, withstand the scrutiny of the defence in a subsequent trial. The trial process will, of course, result in much more expenditure of time and money on the preparation of documents, there is the constant risk of the exposure of sources and methods at some point and, finally, the accused may be acquitted - in 1998 in the UK 43% of the 20,520 contested crown court trials resulted in an acquittal (Barclay & Tavares, 1999, 38).

Now, while the pressures of cost and convenience can be seen to make disruption a very positive option for law enforcement agencies, there are clear dangers involved. The key point about the criminal trial is that it provides some public accounting by the authorities for both the process and substance of their case: their methods of gathering evidence and whether it can establish 'beyond reasonable doubt' the guilt of the accused. Disruption relieves agencies from this need to account and means that those who are subjected to state activities will have no arena in which to challenge what was done to them.

One of the key dangers is the susceptibility of intelligence processes to corruption. To the extent that the defining characteristic of intelligence work is an element of secrecy that cannot be exposed to the public gaze then the possibility of corruption is enhanced. Historically, police corruption scandals have often derived from the use of informants, a technique that is now being deployed more widely than ever by police. As we have seen, the potential disclosure of the use of informants (whether it was proper or improper) in court cases gives rise to one form of corruption, that is, their role not being disclosed to prosecution officials or the court but there is at least some possibility of a check on police procedures. Where disruption is used, however, there is no possibility of this public accounting and the temptations to mis-use informants (or, alternatively, the ease with which informants might manipulate police) are even greater. Certainly, the imminence of the incorporation of the ECHR into law in the UK *via* the Human Rights Act has led to a blizzard of codes of practice and legislation such as the Regulation of Investigatory Practices Bill that are intended to bring procedures into line with due process but the impact of these measures will depend crucially on the interaction of two variables: the extent to which they are translated into practice *via* incorporation into a changing organisational culture and whether these are reinforced by adequate procedures for monitoring and accountability. Other things being equal, the lack of either of these will increase the chances of the new legal edifice representing just a shell of propriety rather than a new age of policing focused on human rights.

Organising intelligence

Before, finally, discussing appropriate accountability procedures, some conclusions about law enforcement intelligence structures are required. Several significant themes emerge from this study: the fact that networks are a better way of describing the contemporary organisation of illegal activities, though some of the nodes of such networks may be more or less formal 'family' or corporate hierarchies; that states struggle to regulate non-state networks (whether legal, illegal or, often, containing elements of both); that hierarchically-organised and a more (USA) or less (UK) fragmented law enforcement community faces particular problems in matching the dynamism of criminal entrepreneurs; and, yet, the intended combination of ICT with strategies of intelligence-led policing is intended to enable the law enforcement community to achieve better regulation of illegal markets.

There is no doubt that the logic of these developments will take law enforcement further away from its traditional para-military, hierarchical roots. If, as its proponents suggest, ILP becomes the organising rationale throughout policing, then this only makes sense within organisational structures that are de-layered and decentered more systematically than has yet happened in the UK. The serious implementation of holistic forms of ILP or problem-oriented policing anywhere requires the control of resources, and therefore organisational interests, to be located at that level of policing appropriate to focus on the problem. Consequently, those 'problems' occurring in some location or 'space' between organisational levels will inevitably be ignored unless some cross-border or multi-agency 'bridge' is constructed. These bridges will be constructed if and when political or other organisational interests come together in sufficient strength to allocate the required resources - there is simply no point in trying to deal with this problem by proposing endless organisational change based on traditional ideas of central co-ordination and amalgamation.

It is possible to identify a number of conditions under which it is more rather than less likely that different agencies (or sections within agencies) will co-ordinate their activities. First, there is the coincidence of interest between agencies. Beyond the simplistic rhetoric of 'all being on the same side', different agencies have varied legal mandates and, given their extensive discretion to identify priorities, may well have even more varied short- to medium-term organisational goals. For example, enforcing the law, reducing social and environmental damage and tackling tax evasion may all make desirable a multi-agency clamp down against ice cream vendors who are selling bootleg cigarettes and beer to children. But sometimes one or another agency might just feel it faces more immediate or serious problems and this happy coincidence of interests may not exist.

Second, co-operation is more likely if previous contacts have produced trust between those involved. People do not like passing on the information they have, for a mixture of positive and negative reasons: they will not pass on information if they worry about unpredictable consequences of their loss of control over it or they may just fear someone else getting the credit for an operation. Security concerns may be real or exaggerated but they will increase the more extensive the network over which the information will be dispersed. The third condition is feasibility. Given broad mandates and limited budgets, the costs of an operation (whether aimed at 'intelligence' or an investigation) will be a prime consideration with a number of contributing variables. The more specific and credible information is, the more likely it is to lead to further commitment of resources. The greater the complexity of a case in terms of jurisdictions and agencies involved, the more likely it is that a formal agreement will need to be negotiated between the contributing agencies, identifying who will do what and, where forfeiture is anticipated, who will get what.

There are also a variety of legal and political conditions that will provide the context for collaboration. A high profile for the 'problem' in politics and the media will increase the pressure on organisations to commit the resources required for collaboration and, where different legal frameworks are in existence, action might be taken within the context of that seen as most permissive by the agencies. This is a consideration in the US with its parallel legal structures; it is not in the UK but it is a consideration in the negotiation of cross-national operations. There is no evidence that establishing overarching organisations with a mandate to co-ordinate the operation of otherwise fragmented agencies actually works. It is more likely that they establish their own interests and form an additional 'level' at which law enforcement operates. Traditionally organised in hierarchical bureaucracies, law enforcement agencies have been prone to 'mirror image' their target and perceive criminal entrepreneurs to be organised in the same way. From the point of view of having an impact on criminal networks, the challenge for law enforcement intelligence is to embrace the network form as, informally, police officers have been doing for a very long time.

Conclusion: making intelligence accountable

But this raises the most profound issues of accountability. There are different ways in which law enforcement agencies might be held accountable. To the extent that their activities are challenged in the trial process, then the courts may hold them *legally* accountable but, as we have seen, the use of many intelligence techniques will not be discussed in court. Second, the fiscal crisis faced by democratic capitalist states in the last quarter of the twentieth century was met in part by the adoption of 'new public management' techniques that sought to apply to state agencies the financial and performance disciplines of the private sector. But this *market* form of accountability is relevant only to notions of efficiency and performance and the idea of people as consumers of police 'services'; it makes little or no contribution to accountability in terms of the fairness, propriety or ethics of police behaviour.

For that, what is required is something that can examine law enforcement from the joint perspective of human rights on the one hand and democratic preferences, on the other. These two, of course, will not necessarily always coincide: classically, when majority preferences are at odds with minority rights. Since the new Human Rights Act will draw the UK courts into making judgments that seek to achieve these sometimes contradictory goals (as US courts have always done and Canadian courts have for the last twenty years) can it be argued that this job should be left up to them? The answer is no - the courts deal with such a small proportion of the 'outcomes' of law enforcement activity, that this cannot be relied upon to provide more than the occasionally significant precedent. The only other possible source is some form of *political* accountability.

This idea gave rise to vigorous debate in the UK in the early 1980s as Labour-controlled local police authorities sought to invigorate their role in order to exercise greater influence over policing policies. The effort was squashed by the Thatcher government abolishing the relevant authorities in part because of their criticisms of police but also because of their pursuit of other policies that were opposed to those of the central government. Thereafter, the authorities were weakened by the reduction of their democratic base and they were drawn into the then fashionable notions of market accountability, performance indicators and all the other paraphernalia of new public management. While much of this has been retained by the Labour government elected in 1997, in other respects the possibility for local authorities in general and police authorities in particular to play a more significant role in matters of crime and policing has been rejuvenated by the Crime and Disorder Act.

However, this still does not provide a framework of accountability that is appropriate to the 'local security networks' of which the public police are just one part. Ian Loader (2000) has recently addressed this issue and suggests the creation of national, regional and local policing commissions to formulate policies and co-ordinate service delivery across the policing network and to bring to account the agencies that comprise it. However much the state's role in *providing* policing may now be limited, there cannot be any institution other than the state that could assert public interest considerations into its *regulation*. The three dimensions of 'public justice' that would guide the work of the commissions would be: processes of consultation to ensure a voice for all individuals and groups, securing the broadest agreement without sacrificing civil rights (including cultivating a human rights culture within agencies) and seeking to provide a 'fair' allocation of policing resources, avoiding the problems of over-policing of some and the underprotection of others. The principal tasks of the commission would include the setting of policies, targets and policing plans including their integration into broader community safety policies; licensing private security companies in the area and awarding contracts; monitoring, inspection and evaluation.

There is much that is interesting and controversial in Loader's proposals; here, however, it is this final group of 'accountability' tasks that are considered. A number of the monitoring and evaluation tasks envisaged by Loader would be concerned with accountability in terms of performance or effectiveness, that is, the extent to which agencies (public or otherwise contracted) were fulfilling their plans, meeting their targets, contractual obligations and so on. But these would also have a crucial democratic or ethical dimension. In either case, the way in which monitoring is conducted is crucial. Currently, much monitoring is done through the provision of reports and statistical indicators and, inevitably, some of this will remain.

Police forces are inspected by Her Majesty's Inspectors of Constabulary who are attached to the Home Office. In recent years the role of HMIC has been given added weight but it is primarily concerned with efficiency and 'value-for-money' matters. As far as propriety is concerned, there is no systematic monitoring - the Police Complaints Authority supervises the police investigation of complaints but it has not succeeded in winning much public confidence. Lord Scarman (1981) noted the impor-tance for police-community relations of a complaints investigation system that would obtain public support and that such a system should not involve police investigating themselves; almost twenty years later the Macpherson Report into the investigation of the death of Stephen Lawrence repeated the same findings (Macpherson, 1999, para.45.22).

The inadequacies of such piecemeal accountability systems have been demonstrated regularly. Loader's proposed Commissions and the not dissimilar proposals made for the reform of policing in Northern Ireland by the Independent Commission headed by Chris Patten (Independent Commission, 1999) would provide for superior forms of monitoring and accountability. What this study of law enforcement intelligence clearly implies for the UK is that the patchwork of officials and organisations with different review responsibilities is too random, uncertain and inadequate. Now that even more policing takes place beyond the public gaze, seeking its goals through informal means and resisting exposure within the criminal justice process, it becomes even more imperative to provide a regulatory mechanism that can compensate. The groundwork has been laid by the new statutory rules and codes of practice for covert policing but they can not be regarded as adequate in their own. Covert policing is even more prone to 'rule-avoidance' than 'normal' policing and therefore it is imperative that some monitoring mechanism be established.

Monitoring would need to be structured at the main levels at which policing is organised - local, force and national. Nationally, the Regulation of Investigatory Practices Bill provides for the establishment of a Covert Surveillance Commissioner to review covert policing and assist the Tribunal if hearing challenges to police action under the Human Rights Act. But as envisaged, the office has only a narrow remit for judicial review; this office should be empowered like the prisons inspectorate to carry out unannounced inspections as well as routine monitoring. This would clearly require an investigative staff. But, ideally, this would be absorbed into a broader-based national policing commission - possibly constructed on the base of the current service authorities for NCS and NCIS. At force level, the police authority would provide the logical basis for a commission with the necessarily enhanced investigative staff. At local, or area, level, it might not be considered a sensible use of resources to establish separate local police commissions with the concomitant staff and there would be genuine security reasons for limiting the numbers of people monitoring the deployment of covert techniques but, since this is the key level of policing as far as most crime is concerned, there certainly is the need to build on the currently under-developed network of community liaison forums established under the Police and Criminal Evidence Act 1984.

The crucial point is that there be created a direct link of accountability and review from the locality to the force area so that concerns can be addressed, for example, that complaints can be independently investigated and adequate monitoring, including random inspections, be conducted to seek to minimise the misuse of covert intelligence and investigative practices that, while indispensable, may also cause great harm both to those on whom they are used and on those deploying them. Throughout this study, the significance of targeting processes has emerged. This is the mechanism by which law enforcement seeks to bridge the gulf between the large number of crime-like incidents or patterns to which it could theoretically respond and the much smaller number for which it is realistically resourced. It is the enormity of this gulf that provides law enforcement with such extensive discretion and the potential for discrimination or unfairness if targeting practices go unchecked. Examining the fairness and proportionality of targeting must become a major focus for external monitors if law enforcement is to be shifted towards addressing broader issues of social damage and away from rounding up the usual suspects and re-cycling them through the criminal justice process.

More generally, if intelligence-led policing is to become more than a rhetorical justification for traditional policing practices, that is, if it is to become a serious attempt to solve the 'knowledge' problem regarding the causes of social damage, then greater efforts are required, especially at the analytical phase of the process. If, as seems likely, policing continues to become more like 'regulation' and less like traditional 'enforcement', then there is even greater need for it to be subject to processes of review and accountability in order to contribute to the maximisation of police effectiveness in terms of justice and rights.

Bibliography and References

Abadinsky H., 1997, *Organized Crime*, 5th Edition, Nelson-Hall, Chicago.

Abraham H.J., 1982, *Freedom and the Court: civil rights and liberties in the United States*, Oxford University Press, New York.

Ackroyd S. *et al*, 1992, *New Technology and Practical Police Work*, Open University Press, Buckingham.

Albini J.L., 1997, 'Donald Cressey's Contributions to the Study of Organized Crime: an evaluation', in P.J. Ryan and G.E. Rush, *Understanding Organized Crime in Global Perspective*, Sage, London, pp. 16-25.

Amey P. *et al*, 1996, *Development and Evaluation of a Crime Management Model*, Police Research Series Paper 18, Home Office, London.

Anderson M. *et al*, 1995, *Policing the European Union: theory, law and practice*, Clarendon Press, Oxford.

Andreas P., 1996, 'US-Mexico: open markets, closed border', *Foreign Policy*, 103, Summer, pp.51-69.

Andreas P., 1999, 'When Policies Collide: market reform, market prohibition, and the narcotization of the Mexican economy', in H.R. Friman and P. Andreas (eds) *The Illicit Global Economy and State Power*, Rowman and Littlefield Publishers, Lanham, Maryland, pp.125-41.

Armstrong G. and Hobbs D., 1995, 'High Tackles and Professional Fouls: the policing of soccer hooliganism', in C. Fijnaut and G.T. Marx (eds) *Undercover - police surveillance in comparative perspective*, Kluwer Law International, The Hague, pp. 175-94.

Arsenault W., 1999, Chief, Homicide Investigation Unit, Office of New York County District Attorney, interview with February 11, New York.

Audit Commission, 1991, *Reviewing the Organization of Provincial Police Forces*, Police Paper No. 9, HMSO, London.

Audit Commission, 1993, *Helping With Inquiries: tackling crime effectively*, Police Paper, HMSO, London.

Audit Commission, 1996, *Detecting a Change: progress in tackling crime*, Police Paper, HMSO, London.

Baldwin R. *et al* (eds.), 1998, *A Reader on Regulation*, Oxford University Press, Oxford.

Baldwin R. and Kinsey R., 1982, *Police Powers and Politics*, Quartet Books, London.

Barclay G.C. and C. Tavares (eds.), 1999, *Digest 4: Information on the Criminal Justice System in England and Wales*, Home Office Research and Statistics Department, London.

Barton A. and Evans R., 1999, *Proactive Policing on Merseyside*, Police Research Series Paper 105, Home Office, London.

Baumgartner M.P., 1987, 'Utopian Justice: the covert facilitation of white-collar crime', *Journal of Social Issues*, 43(3) pp. 61-9.

Bayley D., 1985, *Patterns of Policing*, Rutgers.

Bayley D., 1994, *Police for the Future*, Oxford University Press, New York.

Bayley D., 1999, 'Policing: the world stage', in R. Mawby (ed.) *Policing Across the World: issues for the twenty first century*, UCL Press, London.

Beare M.E., 1996, *Criminal Conspiracies: organized crime in Canada*, Nelson Canada, Toronto.

Beckett K., 1997, *Making Crime Pay: law and order in contemporary American politics*, Oxford University Press, New York.

Benjaminson P., 1997, *Secret Police: inside the NY City Department of Investigation*, Barricade Books, New York.

Bersten M., 1990, 'Defining Organised Crime in Australia and the USA', *Australia and New Zealand Journal of Criminology*, 23, March, pp.39-59.

Bigo D., 1994, 'The European Internal Security Field', in M. Anderson and M. den Boer (eds) *Policing Across National Boundaries*, Pinter, London, pp. 161-73.

BJA, 1998, *The Statewide Intelligence Systems Program*, monograph, Bureau of Justice Assistance Grant # 95-DD-BX-0087, Department of Justice, Washington DC.

Blaney A., 1998, 'Information Gap', *Police Review*, December 4, pp.16-17.

Bliss A. and Lowton P., 1995, *"Sharpening the Focus"; scoring organised crime operations*, SERCS, London.

Block A.A., 1991, *Perspectives on Organized Crime: essays in opposition*, Kluwer Academic.

Bok S., 1986, *Secrets: on the ethics of concealment and revelation*, Oxford University Press, Oxford.

Bottomley K. *et al*, 1991, *The Impact of PACE: policing in a northern force*, University of Hull, Hull.

Bradley D. *et al*, 1986, *Managing the Police: law, organisation and democracy*, Wheatsheaf, Brighton.

Braithwaite J. *et al*, 1987, 'Covert Facilitation and Crime: restoring balance to the entrapment debate', pp. 5-41; 'Overt observations on covert facilitation: a reply to the commentators', pp. 101-22, *Journal of Social Issues*, 43(3).

Bratton W. and Knoblach P., 1998, *Turnaround: how America's top cop reversed the crime epidemic*, Random House, New York.

Bresler F., 1993, *Interpol*, Penguin, Toronto.

Brodeur J.-P., 1985, *On Evaluating Threats to the National Security of Canada and to the Civil Rights of Canadians*, Paper presented to SIRC Seminar, October, pp.7-9.

Brodeur J.-P., 1992, 'Undercover Policing in Canada: wanting what is wrong', *Crime, Law and Social Change*, 18(10-2), September, pp.105-36.

Bunyan T., 1976, *The History and Practice of the Political Police in Britain*, Quartet, London.

Bunyan T., 1993, 'Trevi, Europol and the European State', in T. Bunyan (ed.), *Statewatching the New Europe*, Statewatch, London.

Burke D., 1999, New York City Department of Investigation, interview with, February12, New York.

CACP, 1993, *Organized Crime Committee Report*, Criminal Intelligence Service Canada, Ottawa.

Campbell D. and Connor S., 1986, *On The Record*, Michael Joseph, London, 1986.

Campbell D., 1987, 'ACPO', broadcast on BBC1 as one of the *Secret Society* series.

Campbell D., 1999, Interception Capabilities 2000, www.gn.apc.org/duncan/, May.

Campbell H.J., 1999, Chief Investigator, New York State Ethics Commission, interview with, February 18, Albany.

Card R. and Ward R., 1994, *The Criminal Justice and Public Order Act 1994*, Jordans, Bristol.

CASIS, 1995, *Intelligence Newsletter* #24, Fall, University of New Brunswick, Fredericton, BC.

CASIS, 1998, *Intelligence Newsletter* #33, Fall, Ottawa.

Castells M., 1997, *The Information Age: Economy, Society and Culture, Vol. II, The Power of Identity*, Blackwell, Oxford.

Castleman D., 1999, Chief, Investigation Division, Office of New York County District Attorney, interview with, February 11, New York.

CCRB, 1998, *SemiAnnual Status Report January-June 1998*, VI(I), NY City Civilian Complaint Review Board.

CISO, 1979, 'Overview of Organized Crime in Ontario: an analytical perspective', unpublished, April.

CISO, 1995, *Prospectus*, Toronto, April.

Clay P., 1997, Director of UK Division, NCIS, interview with February 14, Manchester.

Clay P., 1998, 'Taking Transnational Crime Seriously', *International Journal of Risk, Security and Crime Prevention*, 3(2) pp.93-8.

Clegg, S., 1989, *Frameworks of Power*, Sage, London.

CLEU, undated, *Major Administrative Achievements 1989/90 to 1992/93*, mimeo.

CLEU, 1995a, *Background Information*, Province of British Columbia, Ministry of Attorney General.

CLEU, 1995b, *Program Overview 1994*, Ministry of Attorney General, British Columbia.

Cohen A.K., 1977, 'The Concept of Criminal Organisation', *British Journal of Criminology*, 17(2), April, pp. 97-111.

Cook D., 1991, 'Investigating Tax and Supplementary Benefit Fraud', in R. Reiner and M. Cross (eds.) *Beyond Law and Order*, Macmillan, Basingstoke, pp.107-19.

Condon P., 1993, interview with in *Police Review*, April 2.

Constantine T., 1999, *International Drug Trafficking: Law Enforcement Challenges for the Next Century*, speech delivered to the International Conference for Criminal Intelligence Analysts, London, March 1.

Cooper P. and Murphy J., 1997, 'Ethical Approaches for Police Officers when Working with Informants in the Development of Criminal Intelligence in the United Kingdom, *Journal of Social Policy*, 26(1), pp. 1-20.

Cox B. *et al*, 1977, *The Fall of Scotland Yard*, Penguin, Harmondsworth.

Customs, 1999, www.customs.ustreas.gov May.

Dandeker C., 1990, *Surveillance, Power and Modernity*, Polity Press, Oxford.

DCJS, 1996, *District Attorney Organizational and Staffing Survey 1996*, NY State Division of Criminal Justice Services, December.

DCJS, 1998, *Sealing, Expunging and Suppressing in NY State: the role of the sealing Committee*, September.

DeBlock B., 1999, Inspector New York State Police Crime Analysis Unit, interview with, Feb 18, Albany.

den Boer M., 1999, 'Internationalization: a challenge to police organizations in Europe', in R. Mawby (ed.) *Policing Across the World*, UCL Press, London, pp.59-74.

De Roos T., 1998, 'Special police methods of investigation: new legislation in the Netherlands', in Field and Pelser (eds.) *Invading the Private: state accountability and new investigative methods in Europe*, Ashgate, Aldershot, pp.95-110.

De Sousa Santos B., 1992, 'State, Law and Community in the World System: an introduction', *Social and Legal Studies*, 1(2), June, pp.131-42.

Deutsch K, 1966, *The Nerves of Government: models of political communication and control*, Free Press, New York.

Dillon M., 1990, *The Dirty War*, Arrow Books, London.

Dintino J. and Martens F.T., 1983, *Police Intelligence systems in Crime Control: maintaining a delicate balance in a liberal democracy*, C. Thomas, Springfield, Illinois.

Dixon D., 1997, *Law in Policing: legal regulation and police practices*, Clarendon Press, Oxford.

DLE, 1998, *Annual Report 1997-98*, New York State Environmental Conservation Police, Albany.

DOI, 1993, *Report to the Mayor 1990-93*, Department of Investigation, New York.

Donner F.J., 1981, *The Age of Surveillance: the aims and methods of America's political intelligence system*, Vintage, New York.

Donner F.J., 1990, *Protectors of Privilege: Red Squads and police repression in urban America*, University of California Press, Berkeley.

Dorn N. *et al*, 1992, *Traffickers*, Routledge, London.

Dorn N., 1994, 'Three Faces of Police Referral: welfare, justice and business perspectives on multi-agency work with drug arrestees', *Policing and Society*, 4, pp. 13-34.

Dowding K., 1995, 'Model or Metaphor? A critical review of the policy network approach', *Political Studies*, 43(2), pp.136-58.

Dunleavy P. and O'Leary, 1987, *Theories of the state: the politics of liberal democracy*, Macmillan, Basingstoke.

Dunnighan C. and Norris C., 1996, 'A Risky Business: the recruitment and running of informers by English police officers', *Police Studies*, 19(2), pp.1-25.

Dunsire A., 1993, 'Modes of Governance', in J. Kooiman (ed.), *Modern Governance: new government-society interactions*, Sage, London.

Edelman M., 1964, *The Symbolic Uses of Politics*, University of Illinois Press, Urbana.

Edwards A. and Gill P., 1999, *Coming to terms with 'Transnational Organised Crime'*, Paper given to British Criminology Conference, July, Liverpool.

Elliff J.T., 1979, *The Reform of FBI Intelligence Operations*, Princeton University Press, Princeton.

Emsley C., 1996, *The English Police: a political and social history*, 2nd edn., Longman, London.

Engstad P., 1996, Director, Policy Analysis Division, CLEU, interview with, Vancouver, June 14.

Ericson R.V., 1981, *Making Crime: a study of detective work*, Butterworths, Toronto, 1993 edition.

Ericson R.V., 1994, 'The Division of Expert Knowledge in Policing and Security', *British Journal of Sociology*, 45(2) June, pp.151-56.

Ericson R.V. and Haggerty K.D., 1997, *Policing the Risk Society*, Clarendon Press, Oxford.

Fahlman R.C., 1995, OIC, RCMP Criminal Analysis Branch, interview with, May 30, Ottawa.

Farson S., 1991, 'Criminal Intelligence versus Security Intelligence: a re-evaluation of the police role in the response to terrorism', in D.A. Charters (ed.)

Democratic Responses to International Terrorism, Transnational Publishers, New York, pp.191-226.

Faul L., 1995, Intelligence Liaison Officer, CISO, interview with June 21, Toronto.

Field S., 1998, 'Invading the Private? Towards Conclusions', in S. Field and C. Pelser (eds) *Invading the Private: state accountability and new investigative methods in Europe*, Ashgate, Aldershot.

Field S. and Jörg, 1998, 'Judicial Regulation of Covert and Proactive Policing in the Netherlands and England and Wales', in S. Field and C. Pelser (eds) *Invading the Private: state accountability and new investigative methods in Europe*, Ashgate, Aldershot, pp.332-45.

Field S. and Pelser C., 1998, 'Introduction', in Field and Pelser (eds) *Invading the Private: state accountability and new investigative methods in Europe*, Ashgate, Aldershot, pp. 1-28.

Fijnaut C. and Marx G.T., 1995, 'Introduction: the normalization of undercover policing in the West; historical and contemporary perspectives', in Fijnaut and Marx (eds) *Undercover - police surveillance in comparative perspective*, Kluwer Law, The Hague, pp. 1-28.

Fiorentini G. and Peltzman S., 1995, 'Introduction', in Fiorentini and Peltzman (eds) *The Economics of Organized Crime*, Cambridge U.P., Cambridge.

Fitzgerald M., 1999, *Searches in London*, Final Report, mimeo.

Fitzgerald P. and Leopold M., 1987, *Stranger on the Line: the secret history of phone tapping*, Bodley Head, London.

Foucault M., 1983, 'Power, Sovereignty and Discipline', in D Held *et al* (eds), *States and Societies*, Martin Robertson, Oxford.

Foucault M., 1991, 'Governmentality', in G. Burchell *et al* (eds), *The Foucault Effect: studies in governmentality*, Harvester Wheatsheaf, London.

Francis J., 1993, *The Politics of Regulation: a comparative perspective*, Blackwell, Oxford.

Friedrichs D.O., 1995, 'State Crime or Governmental Crime: making sense of the conceptual confusion', in J.I. Ross (ed.) *Controlling State Crime: an introduction*, Garland, New York.

Fuchter, M.,1998, Customs and Excise, Presentation to International Conference for Criminal Intelligence Analysts, Manchester, March 19.

Gallagher J.R., 1996, 'RISS Projects and their impact on analysis over the last decade', *IALEIA Journal* 10(1), September, pp.23-7.

Gamble A., 1988, *The Free Economy and the Strong State*, Macmillan, Basingstoke.

GAO, 1998, *Money Laundering: FinCEN's Law Enforcement Support Role is Evolving*, GGD-98-117, Washington DC, June 19.

Garland D., 1996, 'The Limits of the Sovereign State', *British Journal of Criminology*, 36(4) pp.445-71.

Geraghty T., 1998, *The Irish War*, HarperCollins, London.

Giddens A., 1985, *The Nation-State and Violence*, University of California Press, Berkeley.

Gill P., 1994, *Policing Politics: security intelligence and the liberal democratic state*, Frank Cass, London.

Gill P., 1996, 'Sack the Spooks: do we need an internal security apparatus' in L. Panitch (ed.), *Are There Alternatives?*, Socialist Register, Merlin Press, London, pp.189-211.

Gilligan J., 1999, *The Role of Interpol in the fight against transnational organised crime*, Paper presented to the ESRC Seminar Series on 'Policy Responses to Transnational Organised Crime', Leicester, September.

Goddard S., 1999, 'From Riches to Rags', *Nexus*, 6, Spring, pp.9-11.

Godfrey Jr. E. D. and Harris R., 1971, *Basic Elements of Intelligence: a manual of theory, structure and procedures for use by law enforcement agencies against organized crime*, LEAA, Department of Justice, Washington DC, November.

Godson R., 1994, 'The Crisis of Governance: devising strategy to counter international organized crime', *Terrorism and Political Violence*, 6(2), Summer, pp.163-77.

Goldfarb R., 1995, *Perfect Villains, Imperfect Heroes: Robert F. Kennedy's war against organized crime*, Random House, New York.

Goldstein H., 1990, *Problem-Oriented Policing*, McGraw-Hill, New York.

Gosling J., 1959, *The Ghost Squad*, Doubleday, New York.

Greer S. and South N., 1998, 'The criminal informant: police management, supervision and control', in S. Field and C. Pelser (eds) *Invading the Private: state accountability and new investigative methods in Europe*, Ashgate, Aldershot, pp.31-46.

Gregory F., 1998, 'There is a Global Crime Problem', *International Journal of Risk, Security and Crime Prevention*, 3(2) April, pp.133-38.

Hall R., 1999, 'More light in a dark corner', *Nexus*, 6, Spring.

Haskin J., 1999, Intelligence Officer, New York State Department of Environmental Conservation, Division of Law Enforcement, interview with, February 19, Albany.

Hay C., 1995, 'Structure and Agency', in D. Marsh and G. Stoker, *Theory and Methods in Political Science*, Macmillan, Basingstoke, pp.189-206.

Hay C., 1998, 'The tangled webs we weave: the discourse, strategy and practice of networking', in D. Marsh (ed.) *Comparing Policy Networks*, Open University Press, Buckingham, pp.33-51.

Heaton R., 2000, 'The Prospects for Intelligence-Led Policing: some historical and quantitative considerations', *Policing and Society*, 9, pp.337-55.

Hebenton B. and Thomas T., 1995, *Policing Europe: co-operation, conflict and control*, St. Martin's Press, Basingstoke.

Hebenton B. and Thomas T., 1998, 'Transnational Policing Networks', *International Journal of Risk, Security and Crime Prevention*, 3(2), April, pp.99-110.

Heimer O., 1999, Counsel, Rackets Bureau, Office of New York County District Attorney, interview with February 11, New York.

Held D. and McGrew T., 1993, 'Globalization and the Liberal Democratic State', *Government and Opposition*, 28(2), pp.261-88.

Hennings J., 1997, North Wales FIB, interview with January 29, Colwyn Bay.

Hillyard P., 1987, 'The Normalization of Special Powers: from Northern Ireland to Britain', in P. Scraton (ed.) *Law, Order and the Authoritarian State*, Open University Press, Milton Keynes, pp.279-312.

Hillyard P. and Gordon D., 1995, 'Arresting Statistics: the drift to informal justice in England and Wales', *Journal of Law and Society*, 26(4), December, pp. 502-22.

Hirst P. and Thompson G., 1996, *Globalization in Question*, Polity, Cambridge.

HMIC, 1997a, *NCIS, Report of Inspection*, Home Office, London.

HMIC, 1997b, *Policing With Intelligence: criminal intelligence*, Home Office, London.

HMIC, 1999a, *Merseyside Police 1998/99 Inspection*, Home Office, London.

HMIC, 1999b, *Police Integrity: securing and maintaining public confidence*, Home Office, London.

Hobbs D., 1994, 'Professional and Organized Crime in Britain', in M. Maguire *et al* (eds) *The Oxford Handbook of Criminology*, Clarendon Press, Oxford, pp.441-68.

Hobbs D., 1998, 'Going Down the Glocal: the local context of organised crime', *The Howard Journal of Criminal Justice*, 37(4), November, pp. 407-22.

Hoey A. and Topping I., 1998, 'Policing the New Europe - the information deficit', *International Review of Law, Computers and Technology*, 12(3), pp.527-37.

Hogan G. and Walker C., 1989, *Political Violence and the Law in Ireland*, Manchester University Press, Manchester.

Holden-Rhodes J.F. and Lupsha P.A., 1995, 'Horsemen of the Apocalypse: gray area phenomena and the new world disorder', in G.H. Turbiville (ed.) *Global Dimensions of High Intensity Crime and Low Intensity Conflict*, Office of International Justice, University of Illinois at Chicago, pp. 9-28.

Home Affairs, 1994-5, 3rd Report, *Organised Crime*, HC18-I, HMSO, London.

Home Office, 1984, *Guidelines on the Use of Equipment in Police Surveillance Operations*, mimeo.

Home Office, 1986, *Review of the Regional Crime Squads*, Report of the Tripartite Steering Group, November, London.

Home Office, 1995, *Police and Criminal Evidence Act 1984, Codes of Practice*, revised edition, HMSO, London.

Home Office, 1998, *Guidance on Crime and Disorder Reduction Partnerships*, London. www.homeoffice.gov.uk/cdact/index.htm

Home Office, 1999, *Interception of Communications in the United Kingdom: a consultation paper*, Cm4368, June, London.

Hoogenboom B., 1991, 'Grey Policing: a theoretical framework', *Policing and Society*, 2, pp.17-30.

Howe S., 1997, 'Failing Intelligence', *Policing Today*, 3(2), June, pp.23-25.

Hulnick A.S., 1991, 'Controlling Intelligence Estimates', in G. Hastedt (ed.) *Controlling Intelligence*, Frank Cass, London.

Ianni F.A.J. and Reuss-Ianni E., 1990, 'Network Analysis', in P.P. Andrews and M.B. Peterson (eds) *Criminal Intelligence Analysis*, Palmer Enterprises, Loomis, Calif., pp. 67-84.

Independent Commission, 1999, *A New Beginning: Policing in Northern Ireland*, Belfast.

Innes M., 2000, '"Professionalizing" the role of the police informant: the British experience', *Policing and Society*, 9, pp. 357-83.

Interpol, 1998, *Crime Analysis Booklet*, Version 2.1, August.

IOCA Commissioner, 1998, *Report for 1997*, Cm4001, July, HMSO, London.

Jackson R., 1967, *Occupied With Crime*, George Harrap, London.

Jamieson R. *et al*, 1998, 'Economic Liberalization and Cross-Border crime: the North American Free Trade Area and Canada's Border with the USA, Part II', *International Journal of the Sociology of Law*, 26, pp.285-319.

Johnston L., 1992, *The Rebirth of Private Policing*, Routledge, London.

Johnston L., 1996, 'Policing Diversity: the impact of the public-private complex in policing', in F. Leishman *et al*, *Core Issues in Policing*, Longman, Harlow.

Johnston L., 2000, *Policing Britain: risk, security and governance*, Longman, Harlow.

Joint Task Force on Intelligence and Law Enforcement, 1994, *Report to the Attorney General and Director of Central Intelligence*, August, Washington DC.

Jones T. and Newburn T., 1995, 'How Big is the Private Security Sector?' *Policing and Society*, 5, pp. 221-32.

Jones T. and Newburn T., 1997, *Policing After the Act: police governance after the Police and Magistrates' Courts Act 1994*, Policy Studies Institute, London.

Jones T. and Newburn T., 1998, *Private Security and Public Policing*, Clarendon Press, Oxford.

Joubert Ch., 1994, 'Undercover Policing: a comparative study', *European Journal of Crime, Criminal Law and Criminal Justice*, 2(1), pp. 18-38.

JUSTICE, 1998, *Under Surveillance: covert policing and human rights standards*, JUSTICE, London.

Kagan R., 1984, 'On Regulatory Inspectorates and Police', in K. Hawkins and J.M. Thomas (eds) *Enforcing Regulation*, Kluwer-Nijhoff, Boston, pp.37-64.

Karmen A., 1998, 'Murders in New York City', in Karmen (ed.) *Crime and Justice in New York City*, McGraw-Hill, New York, pp.23-39.

Karp A., 1994, 'The Rise of Black and Grey Markets', *Annals of the American Academy of Political and Social Science*, vol. 535, pp.175-89.

Kedzior R., 1995, *A Modern Management Tool for Police Administrators: strategic intelligence analysis*, CISO, Toronto.

Kelly R.J., 1997. 'Trapped in the Folds of Discourse: theorizing about the underworld', in P.J. Ryan and G.E. Rush, *Understanding Organized Crime in Global Perspective*, Sage, London, pp. 39-51.

Kessler R., 1992, *Inside the CIA*, Pocket Books, New York.

Kickert W.J.M. and Koppenjan J.F.M., 1997, 'Public Management and Network Management: an overview', in Kickert *et al* (eds) *Managing Complex Networks: strategies for the public sector*, Sage, London, pp. 35-61.

Kinsey R., 1985, *A Survey of Merseyside Police Officers*, Merseyside County Council.

Kinsey R. *et al*, 1986, *Losing the Fight against Crime*, Blackwell, Oxford.

Kleinig J., 1996, *The Ethics of Policing*, Cambridge University Press, Cambridge.

Klerks P., 1995, 'Covert policing in the Netherlands', in C. Fijnaut and G.T. Marx, *Undercover - police surveillance in comparative perspective*, Kluwer Law International, The Hague, pp.103-40.

Klerks P., 1999, *The network paradigm applied to criminal investigations: theoretical nitpicking or a relevant doctrine for investigators? Recent developments in the Netherlands*, Paper presented to the ESRC Seminar Series 'Policy Responses to Transnational Organised Crime', Leicester, September.

Kooiman J., 1993, 'Findings, speculations and recommendations', in J. Kooiman (ed.) *Modern Governance: new government-society interactions*, Sage, London.

Lacey N., 1994, 'Introduction: Making Sense of Criminal Justice', in Lacey (ed.) *Criminal Justice*, Oxford UP, Oxford.

Lander A., 1998, NCIS, Presentation to International Conference for Criminal Intelligence Analysts, Manchester, March 19.

Lardiere E., 1983, 'The Justiciability and Constitutionality of Political Intelligence Gathering', *UCLA Law Review*, 30(5), pp.976-1051.

Law Reform Commission, 1986, *Electronic Surveillance*, Law Reform Commission of Canada, Ottawa.

Lawrence P., 1996, 'Strategy, Hegemony and Ideology: the role of the intellectuals', *Political Studies*, 44(1) March, pp.44-59.

Lee M., 1995, 'Across the Public-Private Divide? Private policing, grey intelligence and civil actions in local drugs control', *European Journal of Crime, Criminal Law and Criminal Justice*, 3(4), pp.381-94.

Leigh A. *et al*, 1996, *Problem-Oriented Policing: Brit Pop*, Crime Detection and Prevention Series Paper 75, Home Office, London.

Leigh A. *et al*, 1998, *Brit Pop II: problem-oriented policing in practice*, Police Research Series Paper 93, Home Office, London.

Leishman F. *et al*, 1996, 'Reinventing and Restructuring: towards a "new policing order"', in Leishman *et al* (eds) *Core Issues in Policing*, Longman, London, pp.9-25.

Leng R. and Taylor R., 1996, *Criminal Procedure and Investigations Act 1996*, Blackstone Press, London.

Levesque J.-P., 1995, Special Projects, CISC, interview with, Ottawa, May 29.

Levi M., 1995, 'Covert policing and the investigation of "organized fraud": the English experience in international context', in Fijnaut and Marx (eds) *Undercover - police surveillance in comparative perspective*, Kluwer Law, The Hague, pp.195-212.

Levi M., 1997, 'Evaluating the New Policing: attacking the money trail of organized crime', *The Australia and New Zealand Journal of Criminology*, 30(1), pp.1-25.

Levi M., 1998 'Perspectives on "Organised Crime": an overview', *The Howard Journal of Criminal Justice*, 37(4) November, pp.335-45.

Lloyd, 1990, Interception of Communications Act Commissioner, interview, July 10, London.

Loader I., 1997, 'Policing and the Social: questions of symbolic power', *British Journal of Sociology*, 48(1), pp.1-18.

Loader I., 2000, 'Plural Policing and Democratic Governance', *Social and Legal Studies*, forthcoming.

Loveday B., 1996, 'Business as Usual? The New Police Authorities and the Police and Magistrates' Courts Act', *Local Government Studies*, 22(2), pp.22-39.

Loveday B., 1999, 'The Impact of Performance Culture on Criminal Justice Agencies in England and Wales', *International Journal of the Sociology of Law*, 27, pp. 351-77.

Loverso P., 1999, Acting Chief, Northern Region, Metropolitan Transportation Authority Police Department, interview with, February 23, New York.

Lupsha P., 1996, 'Transnational Organized Crime versus the Nation-State', *Transnational Organized Crime*, 2(1) Spring, pp.21-48.

Lyon D., 1994 *The Electronic Eye: the rise of surveillance society*, Polity Press, Oxford.

McCleay E., 1998, 'Policing policy and policy networks in Britain and New Zealand', in D. Marsh (ed.) *Comparing Policy Networks*, Open University Press, Buckingham, pp. 110-31.

McConville M. *et al*, 1991, *The Case for the Prosecution*, Routledge, London.

McDonald D., 1981, Commission of Enquiry Concerning Certain Activities of the RCMP, Second Report, *Freedom and Security Under the Law*, Minister of Supply and Services, Ottawa.

McDowell D., 1996, *Quo Vadis Strategic Intelligence?* Istana Enterprises Pty Ltd., Cooma, NSW.

McGowan K., 1999, Waterfront Commission, interview with, February 16, New York.

McGrew T., 1992a, 'Conceptualizing Global Politics', in McGrew and P. Lewis *et al, Global Politics*, pp.1-28.

McGrew T., 1992b, 'Global Politics in a Transitional Era', in McGrew and P. Lewis *et al, Global Politics*, pp.312-30.

Macleod, R.C., 1994 'The RCMP and the Evolution of Municipal Policing', in Macleod and D.C. Schneiderman, *Police powers in Canada: the evolution and practice of authority*, University of Toronto Press, Toronto, 1994, pp. 44-56.

MacPhee R., 1995, OIC, Criminal Intelligence Branch, RCMP 'E' Division, interview with, June 16, Vancouver.

Macpherson W., 1999, *The Stephen Lawrence Inquiry*, Cm. 4262-I, HMSO, www.official-documents.co.uk/document/cm4262/

Maguire K., 1990, 'The Intelligence War in Northern Ireland', *International Journal of Intelligence and Counterintelligence*, 4(2), pp.145-65.

Maguire M., 2000, 'Policing by Risks and Targets: some dimensions and implications of intelligence-led crime control', *Policing and Society*, 9, pp.315-36.

Maguire M. and John T., 1995, *Intelligence, Surveillance and Informants: integrated approaches*, Police Research Group Crime Detection and Prevention Series No. 64, Home Office, London.

Maguire M. and John T., 1996, 'Covert and Deceptive Policing in England and Wales: issues in regulation and practice', *European Journal of Crime, Criminal Law and Criminal Justice*, 4, pp. 316-33.

Manning P.K., 1992, 'Information Technologies and the Police', in M. Tonry and N. Morris (eds) *Modern Policing*, University of Chicago Press.

Manning P.K. and K Hawkins, 1989, 'Police Decision-Making', in M Weatheritt (ed.) *Police Research*, Avebury, Aldershot.

Marquis G., 1994, 'Power from the Street: the Canadian Municipal Police, in Macleod and Schniederman, *Police powers in Canada: the evolution and practice of authority*, University of Toronto Press, Toronto, pp. 24-43.

Martens F.T., 1990, 'The Intelligence Function', in P.P. Andrews and M.B. Peterson (eds) *Criminal Intelligence Analysis*, Palmer Enterprises, Loomis, Calif., pp. 1-20.

Martinez R.A., 1997, 'Arizona High Intensity Drug Trafficking Area (HIDTA) Intelligence Program', A. Smith (ed.) *Intelligence-Led Policing*, IALEIA, New Jersey, pp.13-15.

Marx G., 1987, 'The Interweaving of Public and Private Police in Undercover Work, in C.D. Shearing and P.C. Stenning (eds) *Private Policing*, Sage, Newbury Park, pp.172-93.

Marx G., 1988, *Undercover: police surveillance in America*, University of California Press, London.

Marx G., 1995, 'When the Guards guard themselves: undercover tactics turned inward', in Fijnaut and Marx (eds) *Undercover - police surveillance in comparative perspective*, Kluwer Law, The Hague, pp.213-33.

Martens F.T., 1990, 'The Intelligence Function', in P. Andrews Jr. and M.B. Peterson, *Criminal Intelligence Analysis*, Palmer, California, pp. 1-20.

Mathieson T., 1999, *On Globalisation of Control; towards an integrated surveillance system in Europe*, Statewatch, London.

Mawby R., 1999a 'Approaches to Comparative Analysis: the impossibility of becoming an expert on everywhere', in Mawby (ed.) *Policing Across the World*, UCL Press, London, pp.13-22.

Mawby R., 1999b, 'Variations on a Theme: the development of professional police in the British Isles and North America', in Mawby (ed.) *Policing Across the World*, UCL Press, London, pp.28-58.

Mayntz, 1993, 'Governing Failures and the Problem of Governability', in Kooiman (ed.) *Modern Governance: new government-society interactions*, Sage, London.

Merseyside Police, 1983, 'Functions and Duties of Each Rank', Merseyside Police, Liverpool, unpublished.

Metropolitan Journal, 1996, 'Undercover Policing', January, Metropolitan Police, London.

Metro Toronto Police, 1994, *Beyond 2000*, Final Report of Restructuring Task Force, December, Toronto.

Metro Toronto Police, 1995, *Intelligence Services: policy and directives*, June, Toronto.

Mollen M., 1994, *Commission to Investigate Allegations of Police Corruption and the Anti-Corruption Procedures of the Police Department, Report*, City of New York, July 7.

Morgan G., 1986, *Images of Organizations*, Sage, London.

Morgan R. and Newburn T., 1997, *The Future of Policing*, Clarendon Press, Oxford.

Morton J., 1995, *Supergrasses and Informers*, Little, Brown, London.

Murray T., 1982, 'The Co-ordinated Law Enforcement Unit: a Canadian strategy against organised crime', *Police Studies*, 5(2), Summer, pp.21-33.

Nadelmann E., 1993, *Cops Across Borders: the internationalization of US criminal law enforcement*, Pennsylvania State University, PA.

Naylor T., 1995, 'From Cold War to Crime War: the search for a new "national security" threat', *Transnational Organized Crime*, 1(4) pp.37-56.

Naylor T., 1997, 'Mafias, Myths and Markets: on the theory and practice of enterprise crime', *Transnational Organized Crime*, 3(3) Autumn, pp.1-45.

NCIS, 1996, *Annual Report for 1995-96*, London.

NCIS, 1998a, *Annual Report for 1997-98*, London.

NCIS, 1998b, *The Sword*, 8, Spring.

NCIS, 1999a, *Use of Informants: code of practice*, www.ncis.co.uk/web/publications/ September.

NCIS, 1999b, *Surveillance: code of practice*, www.ncis.co.uk/web/publications/ September.

NCIS, 1999c, *Undercover Operations: code of practice*, www.ncis.co.uk/web/ publications/ September.

NCIS, 1999d, *Interception of Communications and Accessing Communications Data: code of practice*, www.ncis.co.uk/web/publications/ September.

NCIS, 1999e, *Recording and Dissemination of Intelligence Material: code of practice*, www.ncis.co.uk/web/publications/ September.

NCIS, 1999f, *Declaration on Ethical Standards and Covert Investigative Techniques*, www.ncis.co.uk/web/publications/ September.

NCIS, 1999g, *Annual Report for 1998-99*, London.

NCIS, 2000, *The National Intelligence Model*, London.

NCS, undated, *Service Plan 1998/99*, National Crime Squad Service Authority, London.

NCS, 1999, *Service Plan, 1999/2000*, National Crime Squad Service Authority, London.

Newburn T., 1999, *Understanding and preventing police corruption: lessons from the literature*, Police Research Series Paper 110, Home Office, London.

Nicholls G., 1996, OIC North West Regional Crime Squad, interview with December 16.

Nogala D., 1995, 'The Future Role of Technology in Policing', in J-P. Brodeur, *Comparisons in Policing*, Avebury, Aldershot, pp.191-210.

Norman P., 1998, 'The Terrorist Finance Unit and the Joint Action Group on Organised Crime: new organisational models and investigative strategies to counter "organised crime" in the UK', *The Howard Journal of Criminal Justice*, 37(4), November, pp.375-92.

Norris C. and Dunnighan C., 2000, 'Subterranean Blues: conflict as an unintended consequence of the police use of informers', *Policing and Society*, 9, pp. 385-412.

North Wales Police, 1996, 'Crime Intelligence Strategy and Policy Framework', OPC/09/96, Colwyn Bay.

NYSOCTF, 1989, *Corruption and Racketeering in the New York City Construction Industry*, Final Report to Governor Mario M. Cuomo, December.

NYSOCTF, 1999, interview with official.

NYSP, 1997, *Annual Report*, NYSP, Albany.

OACP, 1995, *Directory of Corporate Security Administrators and Chiefs of Police*, OACP Corporate Security Liaison Committee.

Oates D., 1999, Chief of Intelligence Division, New York City Police Department, telephone interview with, May 20.

Oldfield R.W., 1988, *The Application of Criminal Intelligence Analysis Techniques to Major Crime Investigation: an evaluation study*, Scientific Research and Development Branch 30/88, Home Office, London.

Organized Crime Independent Review Committee, 1998, *British Columbia's Response to Organized Crime*, Report to Attorney General, September.

OSNP, 1997, *1997 Annual Report*, Office of the Special Narcotics Prosecutor for the City of New York.

Ostrom E. *et al*, 1978, *Patterns of Metropolitan Policing*, Ballinger Publishing, Cambridge, Mass.

Pagano C.L. *et al*, 1981, 'Organized Crime Control Efforts: a critical assessment of the past decade', *The Police Chief*, November, pp. 20-25.

Palango P., 1994, *Above the Law*, McClelland and Stewart, Toronto.

Panorama, 1992, *'Dirty War'*, broadcast on BBC1, June 6.

Parker S, 1997, Senior Intelligence Officer, HM Customs and Excise, interview with June 25, Liverpool.

Passas N., 1993, 'Structural Sources of International Crime: policy lessons from the BCCI affair', *Crime, Law and Social Change*, 20, pp.293-309.

Passas N., 1995, *Organized Crime*, Dartmouth, Aldershot.

Passas N. and Blum J.A., 1998, 'Intelligence Services and Undercover Operations: the case of Euromac', in S. Field and C. Pelser (eds) *Invading the Private: state accountability and new investigative methods*, Ashgate, Aldershot, pp. 143-57.

Penrose R., 1996, National Co-ordinator Regional Crime Squads, interview with December 3, London.

Peters G., 1998, 'Policy Networks: myth, metaphor and reality', in D. Marsh (ed.) *Comparing Policy Networks*, Open University Press, Buckingham, pp.21-32

Peterson M.B., 1995, *Applications in Criminal Analysis: a sourcebook*, Greenwood Press, Westport, Conn.

Peterson M.B., 1999, 'The Basics of Intelligence Revisited', mimeo.

Pollitt C. *et al*, 1998, *Decentralising Public Service Management*, Macmillan, Basingstoke.

Porter M., 1996, *Tackling Cross Border Crime*, Crime Detection and Prevention Series Paper 79, Home Office, London.

Poveda T., 1990, *Lawlessness and Reform: the FBI in transition*, Brooks Cole, Pacific Grove, California.

Powell W.W., 1991, 'Neither Market nor Hierarchy: network forms of organization', in G. Thompson *et al* (eds) *Markets, Hierarchies and Networks: the coordination of social life*, Sage, London.

Prunkun H.W. Jr., 1990, *Special Access Required: a practitioners' guide to law enforcement intelligence*, Scarecrow Press, London.

Punch M., 1979, 'The Secret Social Service', in S. Holdaway (ed.) *The British Police*, Edward Arnold, London, pp.102-17.

Rachal P., 1982, *Federal Narcotics Enforcement: reorganization and reform*, Auburn House, Boston.

Raine L.P. and Ciluffo F.J., 1994, *Global Organized Crime: the new empire of evil*, Center for Strategic and International Studies, Washington DC.

Ransom H.H., 1980, 'Being Intelligent about Intelligence Agencies' in *American Political Science Review*, 74(1) March, pp. 141-48.

Rawlinson P., 1997, 'Russian Organized Crime: a brief history', in P. Williams (ed.) *Russian Organized Crime: the new threat?* Frank Cass, London, pp. 28-52.

RCMP, 1991a, *Criminal Intelligence Program: implementation guide*, RCMP, Ottawa.

RCMP, 1991b, *Criminal Intelligence Program, Roles and Functions: HQ Criminal Intelligence Management Steering Committee*, Ottawa.

RCMP, 1994, 'Outlaw Motorcycle Gangs', *Gazette,* 56(3 and 4), Minister of Supply and Services, Ottawa.

RCMP, 1995, *Criminal Intelligence Program Guide*, RCMP, Ottawa.

RCMP, 1995/96, *Review*, www.tbs-sct.gc.ca/rma/database/studies/ Nov 16, 1999.

Read T. and Oldfield D., 1995, *Local Crime Analysis*, Crime Detection and Prevention Series Paper 65, Home Office, London.

Reiner R., 1992, *The Politics of the Police*, 2nd edition, Harvester Wheatsheaf, Brighton.

Reuss-Ianni E. and Ianni F.A.J., 1983, 'Street Cops and Management Cops: the two cultures of Policing', in M. Punch (ed.) *Control in the Police Organization*, MIT Press, Cambridge, Mass., pp. 251-74.

Richelson J.T. and Ball D., 1990, *The Ties That Bind: intelligence cooperation between the UKUSA countries*, 2nd edn., Unwin Hyman, Boston.

Rider B.A.K., 1995, 'Organised Crime in the United Kingdom: a personal perspective', Memorandum submitted to the Home Affairs Committee, *Organised Crime*, Minutes of Evidence and Memoranda, HC Paper 18:2, London, pp. 193-230.

Rimington S., 1995, 'National Security and International Understanding', Lecture to the English-Speaking Union, London, October 4, London.

Rioux J.-F. and Hay R., 1995, 'Security, Foreign Policy and Transnational Organized Crime: a perspective from Canada', *Transnational Organized Crime*, 1(2) pp.173-92.

Robertson G., 1976, *Reluctant Judas: the life and death of the Special Branch informer Kenneth Lennon*, Temple Smith, London.

Robertson G., 1989, *Freedom, the Individual and the Law*, Penguin, London, 6th edition.

Robertson G., 1994, 'Entrapment Evidence: manna from heaven, or fruit of the poisoned tree?', *Criminal Law Review*, pp. 805-16.

Rogovin C.H. and Martens F.T., 1997, The Evil That Men Do', in P.J. Ryan and G.E. Rush, *Understanding Organized Crime in Global Perspective*, Sage, London, pp.26-36.

Rose D., 1996, *In The Name of the Law: the collapse of criminal justice*, Vintage, London.

Rosen R., 1999, Director, Office of Justice Systems Analysis, New York State Division of Criminal Justice Services, interview with, Feb 17, Albany.

Rosenau J.N., 1992, 'Governance, Order, and Change in World Politics', in Rosenau and E-O. Czempiel (eds) *Governance Without Government: order and change in world Politics*, Cambridge University Press, Cambridge.

Royal Commission on Criminal Justice, 1993, *Report*, Cm 2263, HMSO, London.

Ruggiero V., 1998, 'Transnational Criminal Activities: the provision of services in the dirty economies', *International Journal of Risk, Security and Crime Prevention*, 3(2) 121-9.

Ryan P.J., 1995, *Organized Crime: a reference handbook*, ABC-Clio, Santa Barbara.

Ryder C., 1989, *RUC: a force under fire*, Methuen, London.

Said E.W., 1994, *Representations of the Intellectual: the 1993 Reith Lectures*, Vintage, London.

Sandelli R., 1995, OIC, Metro Toronto Police Intelligence Services, interview with, Toronto, June 9.

Sanders A., 1997, 'Criminal Justice: The Development of Criminal Justice Research in Britain', in P.A. Thomas (ed.) *Socio-Legal Studies*, Dartmouth, Aldershot, pp.185-205.

Savage S. *et al*, 1999, *The Bobby Lobby: police pressure groups and criminal justice policy*, Paper presented to British Criminology Conference, Liverpool, July.

Savage S. and Charman S., 'Managing Change', in F. Leishman *et al* (eds) *Core Issues in Policing*, Longman, Harlow, pp.39-53.

Schelling T.C., 1971, 'What is the business of organized crime?', *Emory Law Journal* 20(1) pp. 71-84.

Scott J., 1991, *Social Network Analysis*, Sage, London.

Scott, 1996, *Report of the Inquiry into the Export of Defence Equipment and Dual-Use Goods to Iraq and Related Prosecutions*, HC115, HMSO, London.

Select Committee on Public Accounts, 1999, *HM Customs and Excise: the prevention of drug smuggling*, 15th Report, House of Commons, London.

Sharpe S., 1994, 'Covert Police Operations and the Discretionary Exclusion of Evidence', *Criminal Law Review*, pp. 793-804.

Shearing C.D. and Stenning P.C., 1981, 'Modern Private Security: its growth and implications', in M.Tonry and N.Morris (eds) *Crime and Justice: an annual review of research*, vol.3, University of Chicago Press, Chicago, pp. 193-245.

Shearing C.D., 1996, 'Reinventing Policing: policing as governance', in O. Marenin (ed.) *Policing Change, Changing Police*, Garland, New York, pp. 285-307.

Sheptycki J., 1996, 'Law Enforcement, Justice and Democracy in the Transnational Arena: reflections on the war on drugs', *International Journal of the Sociology of Law*, 24, pp.61-75.

Shulsky A.N., 1991, *Silent Warfare: understanding the world of intelligence*, Brassey's, Washington DC.

Shuy R.W., 1990, 'Tape Recorded Conversations', in P. Andrews and M.B. Peterson (eds.) *Criminal Intelligence Analysis*, Palmer Enterprises, Loomis, Calif., pp. 117-47.

Silver A., 1967, 'The Demand for Order in Civil Society: a review of some themes in the history of urban crime, police, and riot', in D.J. Bordua (ed.) *The Police: six sociological essays*, John Wiley and Sons, New York, pp. 1-24.

Silverman E.B., 1999, *NYPD Battles Crime*, Northeastern University Press, Boston.

SIRC, 1999, *Annual Report for 1998-99: an operational audit of the Canadian Security Intelligence Service*, Minister of Supply and Services, Ottawa.

Skolnick J.H., 1994, *Justice Without Trial: law enforcement in democratic society*, 3rd edn., Macmillan, New York.

Slapper G. and Tombs S., 1999, *Corporate Crime*, Longman, Harlow.

Smith A., 1997, 'Towards Intelligence-Led Policing: the RCMP experience', in A. Smith (ed.) *Intelligence Led Policing*, IALEIA, New Jersey, pp.16-20.

Smith D., 1971, Some things that may be more important to understand about organized crime than Cosa Nostra', *University of Florida Law Review*, XXIV(1) Fall, pp. 1-30.

Smith D., 1980, 'Paragons, Pariahs and Pirates: a spectrum-based theory of enterprise', *Crime and Delinquency*, July, pp.358-86.

Smith D.J., 1983, *Police and People in London: III A Survey of Police Officers*, Policy Studies Institute, London.

Smith M., 1996, *New Cloak, Old Dagger*, Victor Gollancz, London.

Snider L., 1991, 'The Regulatory Dance: understanding reform processes in corporate crime', *International Journal of the Sociology of Law*, 19, pp.209-36.

Sobocienski D., 1995, 'The Use of Tactical Intelligence in the Development of Organized Crime Control Strategies', *Intelligence - into the 21st Century*, Law Enforcement Intelligence Unit, Sacramento, Calif., pp.22-8.

Sparrow M.K., 1994, *Imposing Duties: Government's changing approach to compliance*, Praeger, Westport, Conn.

Spitzer S., 1977, 'Security and Control in Capitalist Societies: the fetishism of security and the secret thereof', in J. Lowman *et al* (eds) *Transcarceration: essays in the sociology of social control*', Gower, Aldershot, pp.43-58.

Stalker J., 1988, *Stalker*, Harrap, London.

Steele R.D., 1995, 'Private Enterprise Intelligence: its potential contribution to national security', *Intelligence and National Security*, 10(4), October, pp.212-28

Stelfox P., 1998, 'Policing Lower Levels of Organised Crime in England and Wales', *The Howard Journal of Criminal Justice*, 37(4), November, pp.393-406.

Stelfox P., 1999, *Transnational Organised Crime: a selective police perspective*, Paper delivered to ESRC Seminar Series 'Policy Responses to Transnational Organised Crime', Leicester, December.

Stern G.M., 1988, *The FBI's Misguided Probe of CISPES*, CNSS Report #111, Center for National Security Studies, Washington DC, June.

Stewart J.G., 1996, 'Intelligence Analysis of Transnational Crime: assessing Canadian preparedness', *Journal of Conflict Studies*, XVI(1), pp.84-99.

Sullivan R.R., 1998, 'The Politics of British Policing in the Thatcher/Major State', *Howard Journal of Criminal Justice,* 37(3), August, pp. 306-18.

Sutton M., 1998, *Handling Stolen Goods and Theft: a market reduction approach*, Police Research Group Paper, Home Office, London.

Theoharis A.G. and Cox J.S., 1989, *The Boss: J. Edgar Hoover and the Great American Inquisition*, Harrap, London.

Thompson G. *et al*, 1991, *Markets, Hierarchies and Networks: the coordination of social life*, Sage, London.

Thony J.-F., 1996, 'Processing Financial Information in Money Laundering Matters: the financial intelligence units', *European Journal of Crime, Criminal Law and Criminal Justice*, 4(3), pp.257-82.

Tilly C., 1985, 'War Making and State Making as Organized Crime', in P. Evans *et al* (eds) *Bringing the State Back In*, Cambridge University Press, Cambridge, pp. 169-91.

Tombs S., 2000, 'Official Statistics and Hidden crime: researching safety crimes', in V. Jupp *et al*, *Doing Criminological Research*, Sage, London, pp.64-81.

Tremblay P. and Rochon C., 1991, 'Police Organizations and Their Use of Knowledge: a grounded research agenda', *Policing and Society*, 1, pp.269-83.

Turk A.T., 1982, *Political Criminality*, Sage, Beverley Hills.

Urban M., 1992, *Big Boys' Rules: the secret struggle against the IRA*, Faber and Faber, London.

Uglow S., 1999, 'Covert Surveillance and the European Convention on Human Rights', *Criminal Law Review*, pp. 287-99.

Uglow S.& Telford V., 1997, *The Police Act 1997*, Jordans, Bristol.

Van Duyne P.C., 1996, 'The phantom and threat of organized crime', *Crime, Law and Social Change*, 24, pp. 341-77.

Van Duyne P.C., 1997, 'Organized Crime, Corruption and Power', *Crime, Law and Social Change*, 26, pp. 201-38.

Van Duyne P.C., 1998, 'Money-Laundering: Pavlov's dog and beyond', *The Howard Journal of Criminal Justice*, 37(4), November, pp.359-74.

Van Maanen J., 1974, 'Working the Street: a developmental view of police behavior', in H. Jacob (ed.) *The Potential for Reform of Criminal Justice*, Sage, Beverley Hills.

Van Outrive L. and Cappelle J., 1995, 'Twenty Years of Undercover Policing in Belgium: the regulation of a risky police practice', in C. Fijnaut and G.T. Marx (eds) *Undercover - police surveillance in comparative perspective*, Kluwer Law International, The Hague, pp.141-54.

Vertinsky L., 1999, *A Law and Economics Approach to Criminal Gangs*, Ashgate, Aldershot.

Voronin Y.A., 1997, 'The Emerging Criminal State: economic and political aspects of organized crime in Russia', in P. Williams (ed.) *Russian Organized Crime: the new threat?* Frank Cass, London, pp.53-62.

Walden K., 1980, *Visions of Order: the Canadian Mounties in symbol and myth*, University of Toronto Press, Toronto.

Walker C., 1992, *The Prevention of Terrorism in British Law*, 2nd Edition, Manchester University Press, Manchester.

Walker N., 1996, 'Defining Core Police Tasks: the neglect of the symbolic dimension', *Policing and Society*, 6, pp.53-71.

Walker S., 1999, *The Police in America: an introduction*, McGraw-Hill, Boston.

Walsh K., 1995, *Public Services and Market Mechanisms: Competition, Contracting and the New Public Management*, Macmillan, Basingstoke.

Walsh M.E., 1982, 'The Co-ordinated Law Enforcement Unit of British Columbia, Canada: a case study', Battelle Human Affairs Research Centre, Seattle, July, mimeo.

Warner G., 1998, 'Transnational Organised Crime and the Secret Agencies', *International Journal of Risk, Security and Crime Prevention*, 3(2), pp.147-49.

Waterfront Commission, 1997, *1996-1997 Annual Report*, Waterfront Commission of New York Harbour.

Webber F., 1993, 'The new Europe: immigration and asylum', T. Bunyan (ed.) *Statewatching the New Europe*, Statewatch, London, pp. 130-41.

Werdmolder H., 1998, 'Moroccan organised Crime in the Netherlands', *International Journal of Risk, Security and Crime Prevention*, 3(2), pp.111-20.

Wesley S., 1997, Deputy Head, National Intelligence Division, HMC&E, interview with October 22, London.

Whitaker R., 1998, 'Refugees: the security dimension', *Citizenship Studies*, 2(3), pp.413-34.

Williams P., 1994, 'Transnational Criminal Organisations and International Security', *Survival* 36(1), Spring, pp. 96-113.

Williams W., 1997, DCI, North Wales Police, interview with February 13, Wrexham.

Wilsnack R.W., 1980, 'Information Control: a conceptual framework for sociological analysis', *Urban Life*, 8(4), pp. 467-99.

Wilson J.Q. and Kelling G.L., 1982, 'Broken Windows: the police and neighborhood safety', *Atlantic Monthly*, 249, March, pp.29-38.

Winslow R.S. and Burke D.W., 1993, *Rogues, Rascals and Heroes: a history of the NY City Department of Investigation 1873 to 1993*, New York City DOI, Corruption Prevention and Management Review Bureau.

Woodiwiss M., 1988, *Crime, Crusades and Corruption: prohibitions in the US 1900-87*, Pinter, London.

Woodiwiss M., 1993, 'Crime's Global Reach', in F. Pearce and M. Woodiwiss (eds) *Global Crime Connections: dynamics and control*, Macmillan, Basingstoke.

World in Action, 1996, broadcast on October 28 and November 4.

Wright A. *et al*, 1993, *Drugs Squads: law enforcement strategies and intelligence in England and Wales*, Police Foundation, London.

Index